Shortcuts

A Concise Guide to Rotary Cutting

Revised and Updated with All-New Quilts

DONNA LYNN THOMAS

Martingale
& COMPANY

Bothell, Washington

Dedication

This book is dedicated to my beloved husband, Terry, and my two fine sons, Joseph and Peter.

Acknowledgments

I'd like to thank everyone who helped make the revision of this book possible.

First of all, a big thank-you to the quilt-makers who test the patterns and supply sample quilts. It is always important for another set of eyes to use and test a pattern for accuracy before it goes into a book. Heartfelt thanks to Dee Glenn, Deb Rose, Linda Kittle, Gabriel Pursell, Robin Chambers, Ann Woodward, Ursula Reikes, and Kari Lane for all their invaluable hard work. Dee and Deb have been with me since my very first book, *Small Talk*, almost ten years ago!

Quilts are flat and relatively lifeless until they are quilted. As always, I am blessed with some of the best quilters around. Ann Woodward, Judy Keller, and Kari Lane add beauty, depth, and life to my quilts. I am beholden to each of you—thanks so much.

A big thanks to my friend Gertrud Hartwig for asking questions and keeping me on my toes!

This is my fifth book with Ursula Reikes as the technical editor. She has become a friend in our shared passion for not just quilting, but gardening. She picks up all the mistakes that get past the pattern testers for me and always finds a better way to explain complicated things. We brainstorm and gnash our teeth together and always seem to have fun doing it. She is a stickler for detail and always makes me look better than I am! Thank you, Ursula—I absolutely cannot imagine doing a book without you.

Credits

President .... Nancy J. Martin
CEO/Publisher .... Daniel J. Martin
Associate Publisher .... Jane Hamada
Editorial Director .... Mary V. Green
Technical Editor .... Ursula Reikes
Design and Production Manager .... Cheryl Stevenson
Cover and Text Designer .... Stan Green
Copy Editor .... Liz McGehee
Illustrator .... Laurel Strand
Photographer .... Brent Kane

Martingale & Company

That Patchwork Place, Inc., is an imprint of Martingale & Company.

Shortcuts: A Concise Guide to Rotary Cutting

Martingale & Company, PO Box 118, Bothell, WA 98041-0118 USA

Printed in Canada
04 03 02 01 00 99 6 5 4 3 2 1

Library of Congress Cataloging-in-Publication Data

Thomas, Donna Lynn,
Shortcuts : a concise guide to rotary cutting / Donna Lynn Thomas.
—Rev. and updated with all-new quilts.
p. cm.
ISBN 1-56477-260-8
1. Patchwork. 2. Rotary cutting. I. Title.
TT835.T45 1999
746.46—dc21 99-10298
CIP

MISSION STATEMENT

We are dedicated to providing quality products and service by working together to inspire creativity and to enrich the lives we touch.

Table of Contents

Preface

The last twenty-five years have been exciting for quiltmakers. In the seventies, quiltmaking underwent a massive revival, where mostly traditional techniques prevailed. In the eighties, all kinds of new ideas and techniques emerged, along with fundamental changes in the way we assembled quilts. About this time, the rotary cutter was introduced, changing quiltmaking forever. It was a thrilling time to be a quiltmaker—on the cutting edge, so to speak!

By 1990, it was obvious rotary cutting was here to stay. An abundance of books dealt with different aspects, ideas, and techniques of rotary cutting, but there was no definitive and comprehensive reference book for quiltmakers who were just purchasing their first rotary cutter—no point A from which to start and grow. And so, the idea for *Shortcuts* was born.

Living and teaching quiltmaking in Germany at the time, I knew there was also a need for a metric version. The original *Shortcuts* was written in the English imperial version (inches) as well as two metric versions, one based on a .5cm seam allowance and one based on a .75cm seam allowance. These metric versions were translated into German (.75cm), Dutch (.5cm), French (.5cm), and Japanese (.75cm). Along with the English metric version (.75cm), there were six published versions of Shortcuts. Since its release in 1991, *Shortcuts* has sold more than 130,000 copies worldwide.

Donna Lynn Thomas

Introduction

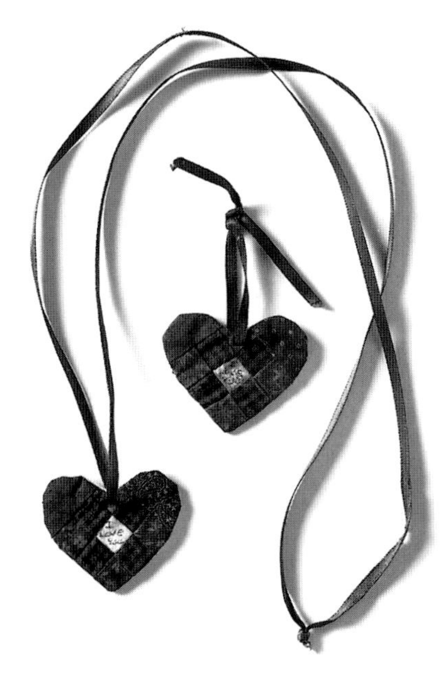

It's hard to believe it's been eight years since Desert Storm, during which I wrote the original *Shortcuts.* My husband, Terry, was part of the ground combat forces there and returned to us in Germany from Kuwait in the summer of 1991. That was a particularly difficult time for me, but as always, life moves on, and he has since retired from the army and started a new civilian life.

Changes have occurred in the world of quiltmaking, too, and it seemed time to update and revise *Shortcuts.* I have completely reorganized the book, adding, removing, and replacing material. In the years since I first wrote *Shortcuts,* bias strip piecing has become more refined and extensive in nature. The very basics of bias strip piecing are still included in the revised *Shortcuts,* but I deal with the subject exclusively and completely in my book *Stripples Strikes Again!*

As it happens, the Bias Stripper ruler I designed to accompany the newer bias strip-piecing processes also makes some basic rotary-cutting chores easier. Therefore, the Bias Stripper is presented in the new *Shortcuts* as an alternative to some methods.

Although the idea of rotary cutting is to avoid the use of templates, there are specific occasions when paper cutting guides are necessary. As you read *Shortcuts,* you will learn how to tape these guides to a rotary ruler so you can quick-cut odd-sized pieces, thereby avoiding laborious template marking and cutting.

You will also find some new and interesting ways to modify units into other shapes without a lot of fussy piecing. I have not included this information in any of my books to date, but these methods simply build on existing concepts.

In some cases, where there is more than one way to achieve the same cutting goal, I have presented options so readers can select their favorites. What works well for one is not always easiest for another.

If you have the original *Shortcuts,* I think you'll find the new information and organization in this updated version interesting and useful. It really is quite different. If this is your first copy, I hope you find it as basic and solid a reference as so many have in the past.

General Information

One of the first things you'll notice about rotary cutting is the absence of a marked sewing line on the fabric. In traditional hand piecing, sewing lines are marked, and seam allowances are added beyond the marked line. In rotary cutting, seam allowances (the traditional ¼" on all sides) are included in the cut dimensions of the pieces. Be sure to remember this when working with patterns and books written for rotary cutting. Unless specifically stated otherwise, the sizes referred to in these patterns are cut sizes, not finished sizes.

Consequently, when planning and adapting your own patterns and ideas to rotary cutting, you must know how much to add to the planned finished size of each piece for seam allowances, to determine the proper cut size. In addition to actual rotary-cutting processes, *Shortcuts* teaches you how to determine cut sizes for your own projects.

All the accurate cutting methods in the world won't mean a thing if your sewing is not accurate. Since precise seam-allowance dimensions are included in the pieces you cut, it is imperative that you sew an accurate ¼" seam. If you sew too wide or too narrow a seam, the small errors on each seam snowball into frustrating, inaccurate piecing. Ideally, your seams and intersections should just fall together into a perfect match.

Conduct a strip test on your machine to test the accuracy of the ¼" seam guide and your ability to use it correctly.

1. Cut 3 strips of fabric, each 1½" x 3". Check the width of each strip for accuracy.
2. Sew the strips together side by side. Align the raw edges carefully, and sew slowly and accurately using the machine's ¼" seam guide.
3. Press the 2 seam allowances away from the center strip. The center strip should measure exactly 1" from seam to seam.

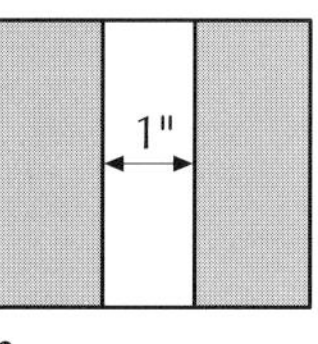

If the center strip is off by just a thread or two, check your sewing habits first. Were the raw edges perfectly aligned and did you keep them that way while stitching? Did you sew too fast to control the edges while stitching? Do you tend to wander when stitching as opposed to sewing a straight seam? Were the strips exactly 1½" wide, or were they just kind of close? These little things are often the source of inaccurate seams. The solution is to slow down. Take the time to be careful and accurate when cutting and sewing.

Your machine could also be the problem. Does your presser foot hold the fabric layers snug enough to keep them aligned? Do the feed dogs feed fabric through without shifting layers? If the machine does not operate properly, get it repaired. The reduction in frustration and seam ripping more than compensates for the effort.

If, despite careful stitching, the center strip does not measure exactly 1" wide, check the guide.

1. Cut a 2" x 6" piece of ¼" graph paper. Put the paper under the presser foot and lower the needle into the paper, just barely to the right of the first ¼" grid line, so that the needle is included in the dimension of the seam allowance. Otherwise, the stitching will decrease the size of the finished area by a needle's width on each seam you sew.
2. Adjust the paper so it runs straight forward from the needle, angling neither to the left nor to the right. Lower the presser foot to

hold the paper in place. Tape the left edge of the paper down so it won't slip.

3. Check the machine's ¼" guide against the edge of the graph paper. If the guide is the edge of the presser foot, the edge should run along the edge of the graph paper. If the guide is an etched line on the throat plate, the same should be true.
4. If the edge of the presser foot or etched line does not run along the edge of the graph paper, you need to make a new guide. Stick a piece of masking tape or adhesive-backed moleskin along the edge of the graph paper as shown. Make sure it is in front of and out of the way of the feed dogs.

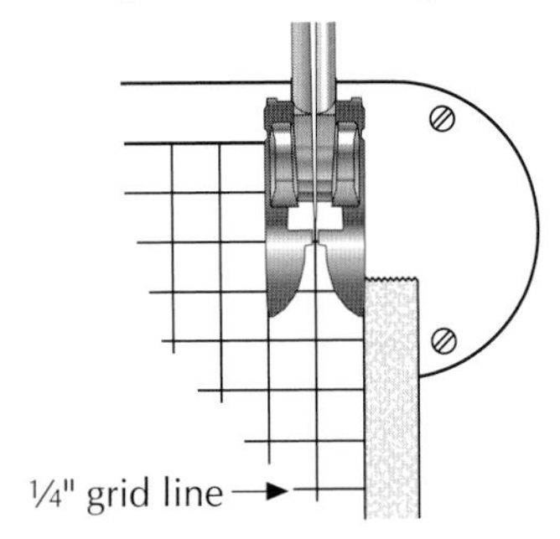

Put masking tape in front of needle along edge of graph paper to guide fabric.

Do another strip test to check this new guide. Adjust the tape guide as necessary until you can conduct strip tests accurately several times in a row. If you are using masking tape, build up the guide with several layers of tape to create a ridge that will help you guide the fabric.

EQUIPMENT

There are a tremendous number of books and patterns written today for rotary cutting. Listed below are the supplies and equipment you will need for working with those patterns or for adapting your own ideas to rotary cutting.

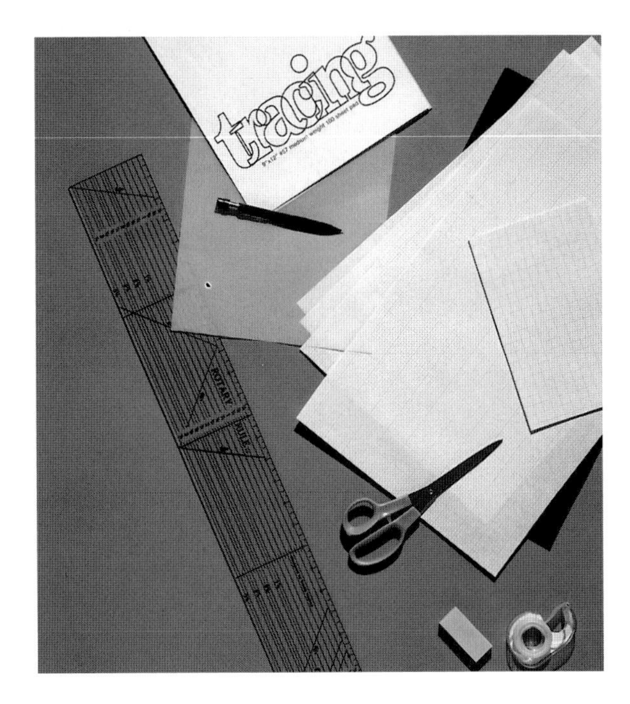

Basic Supplies

Graph paper. Every design begins with a drawing, whether full size or to scale. It is important to have ⅛"-grid graph paper on hand for your drawings. Some people prefer paper with the inch lines highlighted, but it is not necessary.

Lead pencil. A fine-point mechanical pencil is best, but a well-sharpened #2 lead pencil and a sharpener works well.

Eraser. Be sure to use a good-quality eraser for mistakes. An engineer's eraser does not damage the paper as easily as the basic school eraser does.

Tracing paper. Tracing paper is used for making paper cutting guides that are taped to your rotary ruler for cutting odd-sized shapes.

Transparent tape. Use removable tape for your paper cutting guides or they will tear.

Drawing ruler. I use my 3" x 18" rotary ruler for drawing, not only because it is accurate and has all the markings I need, but also because its angle markings are so helpful in drawing diamonds, triangles, and other geometric shapes.

Scissors. You need a sharp pair of paper scissors for cutting out paper cutting guides.

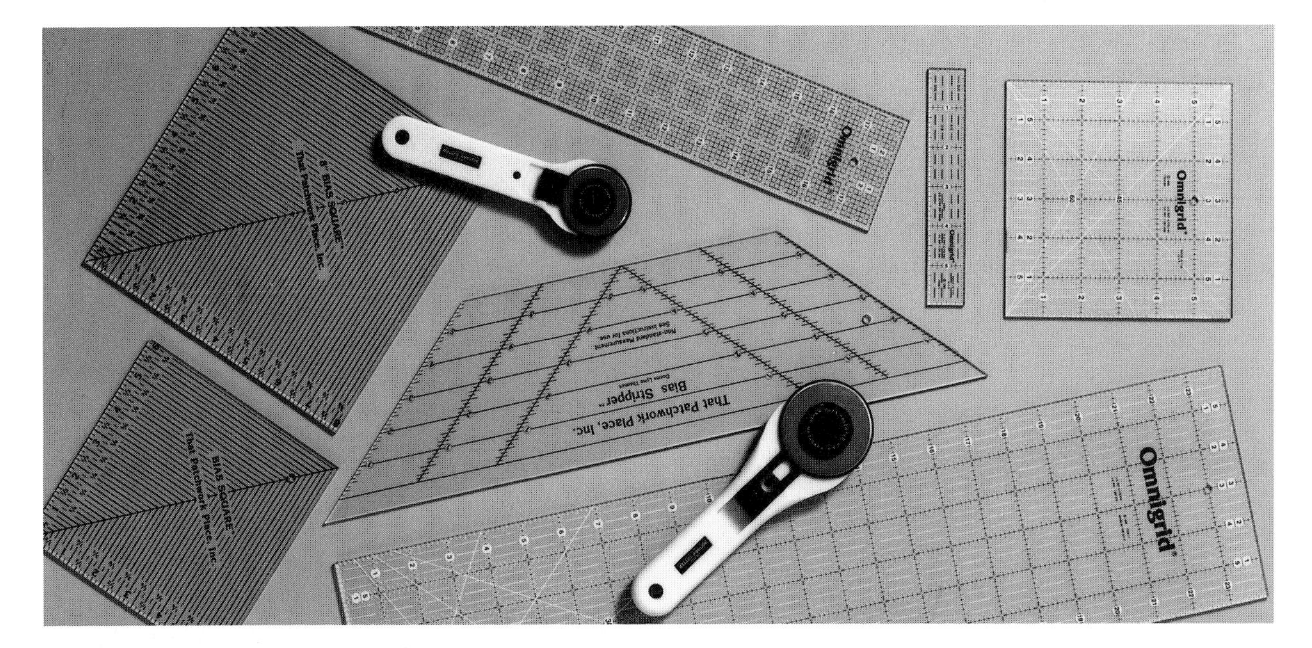

Rotary Equipment

Rotary cutters. A rotary cutter is a cutting instrument with a round blade attached to a handle. It looks like a pizza cutter with a protective shield that is either manually or automatically released, depending on the model. Rotary cutters come in a number of different sizes. I prefer the largest cutter for general cutting, especially when cutting through many layers of fabric. When working with miniature quilts, though, I prefer the small cutter, because it is more maneuverable.

Be sure to keep a replacement blade on hand. Periodically, the blade will become dull or nicked and must be replaced. It is also necessary to remove the lint between the blade and the front sheath from time to time. Dismantle the cutter, carefully wiping the blade free of lint or dirt with a clean, soft cloth. Add a small drop of sewing-machine oil to the part of the blade lying under the sheath. You may find that your cutter feels like it has a new blade after this simple cleaning process.

Rotary blades are extremely sharp cutting instruments and must be treated with a great deal of care to avoid accidents. Please keep these tools well out of the reach of children—they can easily sever tiny fingers.

Before using a cutter with a manual safety shield, tighten the back screw so that the safety shield cannot be easily pushed back from the blade with simple pressure on the cutter. Make it a conscious habit to engage the manual safety mechanism at the completion of every single cutting stroke.

Automatic safety shields have a spring mechanism that is designed to push away and expose the blade when pressure is exerted on the cutter. When pressure is removed from the blade, it springs back to its covered position. In my mind, this is false safety since the shield will just as easily retract if bumped by small hands or if the cutter falls off the table and lands on a foot or leg. I also find the automatic shield reduces your ability to cut through several layers of fabric well.

Always cut away from your body. One slip or overly powerful stroke toward yourself could result in a painful cut on your thigh or elsewhere.

Rotary mats. You must have a special rotary mat to use with your rotary cutter. If you try to cut fabric on anything but a rotary mat, you will immediately ruin both your blade and the cutting surface.

Always store your mats flat and keep them away from extreme hot or cold temperatures that can warp them irreparably. Keep hot items such as irons, coffeepots, or mugs off the mats for the same reason.

There are many different types of mats available today. Some come with measured grid lines and some are double-sided. The measurements are useful for rough cuts only. Precision cutting is best done with your ruler and cutter as discussed later.

Rotary rulers. A good rotary ruler is an invaluable tool and a necessity for rotary cutting. There are many types of rulers on the market today, ranging from highly specialized tools to general-purpose rulers. They come in all shapes and sizes with an assortment of markings. Look for the following features when choosing a rotary ruler.

- A rotary ruler should be made of transparent, ⅛"-thick, hard acrylic.
- A ruler with ⅛" markings is absolutely necessary. These ⅛" marks should appear on every inch line, both horizontally and vertically.
- A ruler with 30°, 45°, and 60° lines is essential. The corner of the ruler is the 90° guide.
- A "window" at the intersection of each inch line is helpful in guaranteeing that the edge of your fabric is where it ought to be in relation to the markings.

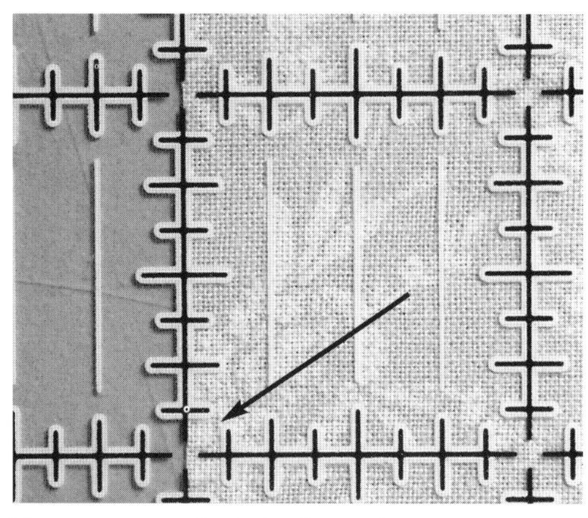

"Window" on ruler

- A 3" x 18" or 6" x 12" ruler is a nice size for most work as long as the fabric is folded twice: from selvage to selvage and again from the fold to the selvages.
- A 24"-long ruler is useful for cutting large strips and shapes and for working with fabric folded only once, from selvage to selvage. These rulers are generally 6" wide, which is helpful when cutting wide border strips.
- A 12" or 15" square ruler is handy for cutting border strips or other pieces that are wider than 6". These large rulers are also great for straightening the edges of an on-point quilt or trimming a long edge and corner before applying binding.
- A 1" x 6" ruler is helpful when working with small quilts or small pieces. I keep one by my sewing machine to constantly check my work for accuracy. It is also the perfect ruler for checking your strip test.
- The Bias Square® ruler is another invaluable ruler. It is a 4", 6", or 8" square with a diagonal line across its middle and measurements on two adjacent sides. Its main function is to cut presewn squares from bias strip units, but it is very useful for other tasks, such as nubbing triangle points or cutting simple squares.
- The Bias Stripper™ ruler, although designed for another purpose, is quite useful for cutting shapes that would normally require more tedious methods. If you have one on hand, you'll be pleased to find new ways to put this ruler to work for you. *Do not use the Bias Stripper ruler for general-purpose rotary cutting; it has nonstandard measurements.*

Getting Ready

Fabric and Grain Line

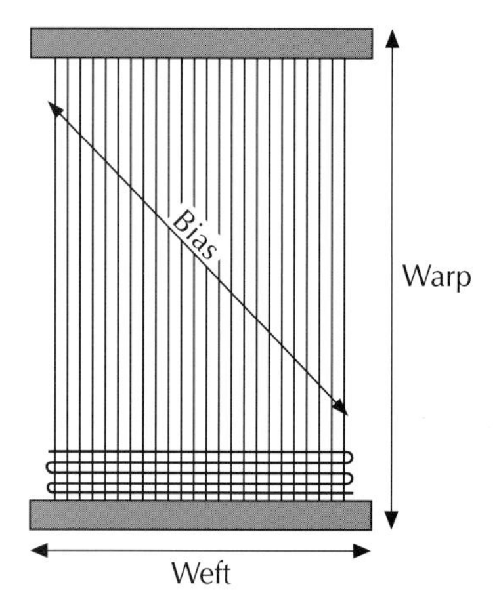

Warp yarns form the lengthwise grain. Weft yarns are woven back and forth to form the crosswise grain.

Understanding the weaving process helps explain why each type of fabric grain has its own special properties. When fabric is woven on a handloom, the process begins with the long warp yarns attached to the front bar of the loom. If ten yards of fabric are desired, then the warp yarns are cut ten yards long, plus enough extra yardage to roll the ends tightly onto the bar at the opposite end of the loom. There are as many yarns lined up across the front bar as are needed to make the fabric the desired width. When the fabric is finished, these warp yarns are referred to as the lengthwise grain of the fabric. The outside yarns in the warp also form the selvages of the fabric.

Once the warp yarns are secured in place, yarns are wound on a shuttle and woven back and forth from side to side through the warp yarns. These side-to-side weft yarns become the crosswise grain of the finished fabric.

Lengthwise grain has little or no give since the warp yarns are tightly secured at both ends during the weaving process. The lack of give means that edges cut parallel to this grain will not stretch with handling.

Crosswise grain has a slight amount of give since the weft yarns are not secured to anything except the warp yarns. Even so, the yarn will give only so far. Edges cut parallel to the crosswise grain can stretch slightly if roughly handled. Generally, cutting on either the lengthwise or crosswise grain is considered to be cutting on-grain. Occasionally, you will find that a pattern specifically instructs you to cut on one type of grain instead of the other.

Bias is anything other than lengthwise or crosswise grain, although true bias is defined as the direction running at a 45° angle to the other grains. Think of the lengthwise and crosswise grains as forming a square. Bias runs from corner to corner across the diagonal of the square. It has a generous amount of give when pulled, since there are no diagonal yarns restraining it. Be careful when handling edges cut parallel to the bias. They can easily become distorted, stretched, and wavy if pulled and handled roughly.

Generally, when making quilts, you should try to cut shapes as close to straight-grain as possible. It's difficult to rotary cut strips that are true straight-grain, so "close" grain usually provides satisfactory results. Due to the quirks of mass production, few fabrics are printed on-grain, and many are stretched off-grain when rolled onto bolts. A piece of fabric that is badly off-grain can sometimes be pulled straight by holding opposite diagonal corners and gently pulling.

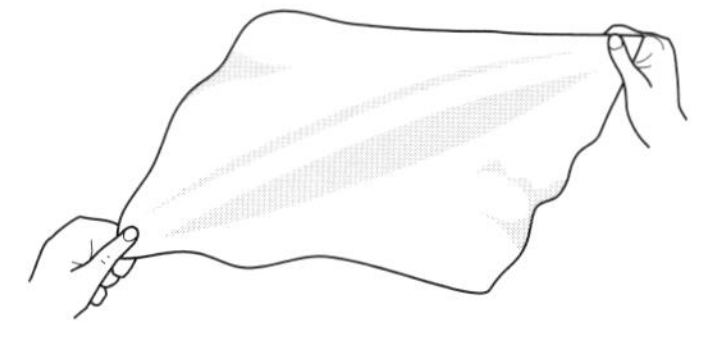

All squares, strips, and rectangles should be cut on-grain. Some shapes, such as triangles, cannot have all edges cut on-grain. Therefore, it is a good idea to look at the position of the shape in the pattern and consider two rules when deciding which edges to cut on-grain.

Rule 1: Place all edges that will go on the perimeter of a quilt block on-grain so the block does not stretch out of shape.

Rule 2: Whenever possible, without violating Rule 1, sew a bias edge to a straight edge to stabilize the seam.

NOTE: The cutting directions in *Shortcuts* are based on the most common placement of grain line. You may need to make your own adjustments for different grain situations.

Fabric Preparation

Prewashing

Before prewashing your fabrics, it's a good idea to check them for bleeding. Soak dark and light fabrics separately in very warm water. If the water is clear after twenty minutes, the fabric is ready for prewashing. If not, rinse and soak again. If the fabric still bleeds after several rinses with no sign of letting up, do not use the fabric.

Once fabrics pass the bleeding test, wash them in warm water with sudsy ammonia (¼ cup for a machine or 1 tablespoon for a sink) or commercial quilt soap. Do not use laundry detergent because it can fade fabrics or cause otherwise stable dyes to bleed. Dry on low to medium heat until damp-dry. Press with a hot iron. Gently straighten and refold your fabric from selvage to selvage.

Making a Cutting Edge

Once your fabric is washed and pressed, you must trim the raw edges before cutting it into the pieces you need.

1. Lay the pressed fabric on the rotary mat, with the fold toward you and the selvages at the top edge of the mat. Place the ragged edges to the left if you are right-handed, and to the right if you are left-handed.
2. Place the edge of your rotary ruler inside the raw edge of the fabric. To make a cut at a right angle to the fold, lay one edge of the Bias Square along the fold of the fabric and adjust the straight ruler so that it is flush with the Bias Square.

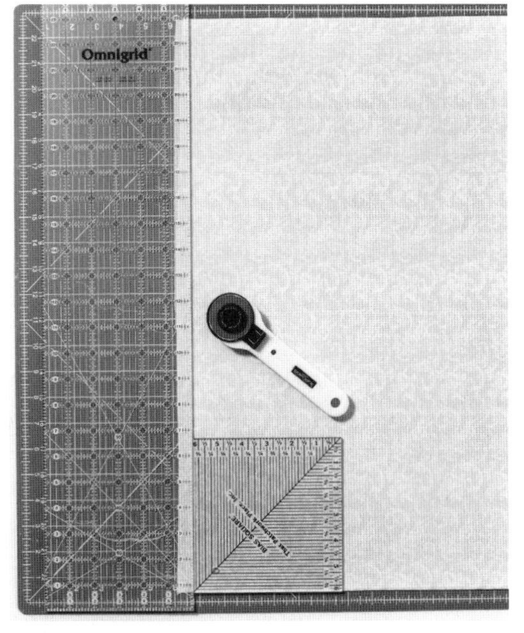

Right-handed cutting

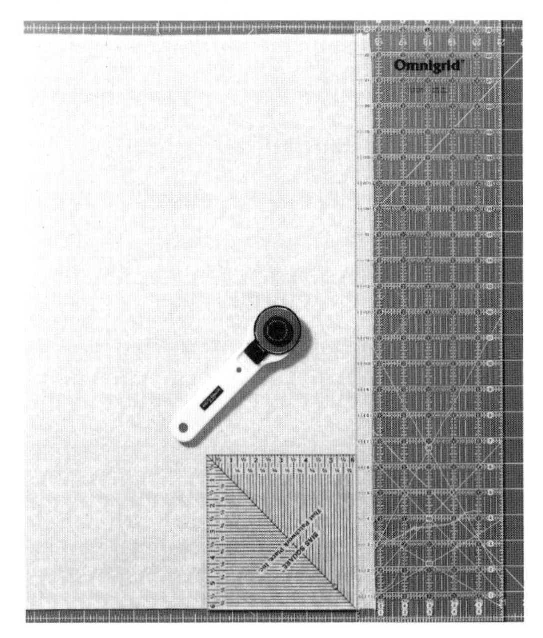

Left-handed cutting

3. Hold the ruler with firm, downward pressure and your fingers spread wide so the ruler doesn't shift. Some quilters find it helpful to anchor their ruler by placing their outer fingers or the palm of their hand to the side of the ruler away from the cutter.
4. Push the Bias Square out of the way. Retract the rotary cutter's safety mechanism, place the blade next to the ruler's edge, and begin to cut slowly away from yourself with firm, downward pressure. As the blade rolls along the ruler's edge, you may need to slowly and carefully "walk" your hand up the ruler. Be careful not to shift the ruler out of line. Cut completely past the selvages and engage the safety mechanism before putting the cutter back on the table. This newly cut edge is your straight-of-grain cutting edge.

Some quilters prefer to fold their fabric one more time, bringing the fold up to match the selvages before cutting. Some of us with short arms or arthritis lose our control and power over the longer distance of the single fold. We have better control over the shorter distance of the double fold. Others have only a 12"-long ruler that can't span the longer distance of the single fold. Some do not have a mat large enough to handle the single-fold fabric. If any of these situations apply to you, by all means fold your fabric a second time. However, you must exercise extra care when cutting it.

The basic process of making a cutting edge remains the same except for the addition of the second fold. When making the second fold, be sure to press the fabric both before and after folding. You must ensure that the layers in the fold are neatly and closely aligned, to avoid creating a "bent" strip later. For the same reason, it is also imperative to make each and every cut at a right angle to the folds.

Cutting Strips, Squares, and Rectangles

Strips

Almost all rotary cutting begins with strips of fabric that are then cut into other shapes, such as squares, triangles, rectangles, and diamonds. If the strips will be cut into simple squares, rectangles, or used "as cut" for border strips, they are cut ½" wider than the desired finished size. If they are to be crosscut into triangles, rectangles, or diamonds, the strips are cut the required cut sizes of those particular shapes.

Once you have a clean-cut edge, you can cut strips accurately. If you are right-handed, measure in from the right edge of the ruler to the desired strip width. Align this measurement with the clean, left edge of the fabric. In addition, align one of the ruler's horizontal lines on the fold at the bottom of the fabric. If the cut is not at a right angle to the folds, you will end up with a "bent" strip when you unfold it. If you are left-handed, measure in from the left side of the ruler and align the measurement with the right edge of the fabric. Cut strips from the bottom to the top in the same fashion as when making the initial cutting edge.

NOTE: Left-handed quilters will always cut everything from right to left, not from left to right as right-handed quilters do. Do not try to fight yourself and work in a way that is inappropriate for your body. You will only end up with inaccuracies and awkward, low-power cuts. Remember to reverse the instructions as you do for all other instructions. Use a mirror, if necessary, to better visualize the reversal.

To cut a strip to a particular length, first trim away the selvages at the ends of the strip. Then measure in the desired distance from the end of the strip and cut to size. Be sure to have one edge of the ruler aligned with the long cut edge of the strip so the crosscut is at a right angle to the edge of the strip.

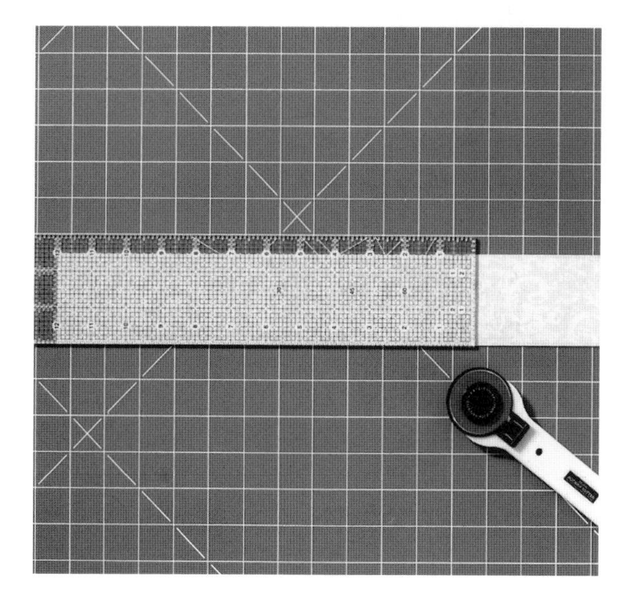

TIP

Ruler lines are often thick, and many people wonder on which side of the line they should cut. The answer is to position the edge of the fabric so it runs through the center of the line. This is where rulers with crosshair windows have an added advantage. You can see the edge of the fabric as it runs through the center of the ruler line.

Squares

Cut squares from strips of fabric. First, determine the cut size of the square by adding ½" to the desired finished size. Cut strip(s) this width. Remove the selvages with right-angle cuts. In the same fashion as cutting strips to length, crosscut the strip(s) into squares. Again, always be sure to place a ruler line on the long edge of the strips when cutting. If, after several cuts, you cannot align the ruler lines on both the long and short edges of the strip at the same time, trim the short edge at a right angle to the long edge again.

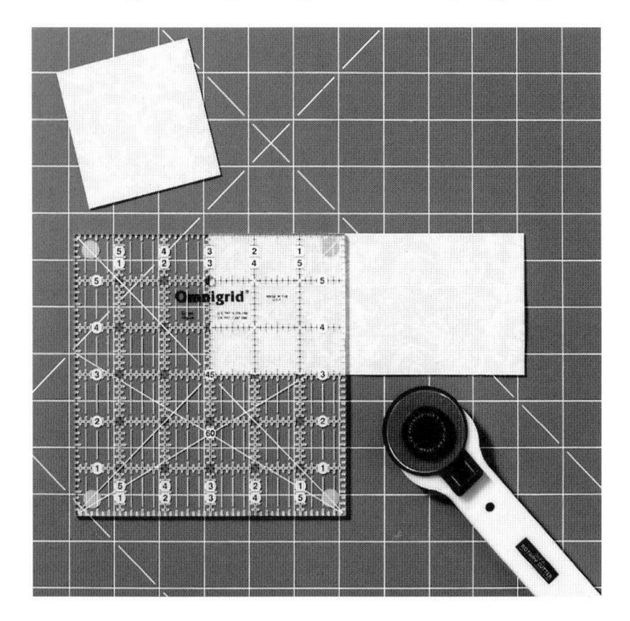

Rectangles and Bars

Rectangles and bars are cut in the same fashion as squares. A true finished-size rectangle is twice as long as it is wide, such as 2" x 4", 3" x 6", or 2¾" x 5½"; anything else, such as 2" x 3" or 2" x 5", is considered a bar. The cut size of each is always ½" larger than the finished size on both dimensions.

NOTE: The cut size of a true rectangle will not reflect the 2-to-1 ratio found in the finished size. For instance, a 2" x 4" finished size rectangle will be cut at 2½" x 4½". The cut size of the length (4½") is not twice the width of the cut size (2½"). Be sure to look at finished sizes, not cut sizes, when determining if a rectangle is a true rectangle.

Cut strips the cut width of the rectangle or bar and then crosscut the strips into units at the desired cut length.

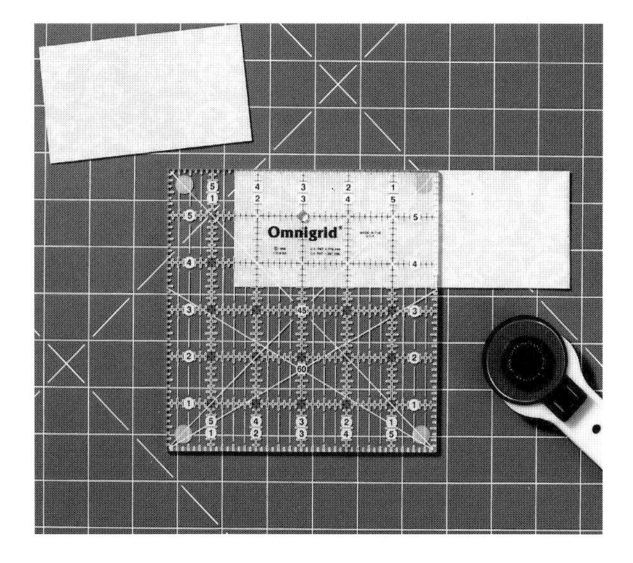

Cutting Right Triangles

There are two types of right triangles, which are created by cutting squares differently: half-square triangles and quarter-square triangles. Depending on whether you cut the squares in half or in quarters, you can alter the position of the triangles' straight grain.

Half-Square Triangles

Make half-square triangles by cutting a square in half once diagonally to yield two right triangles with the straight grain on the two short edges.

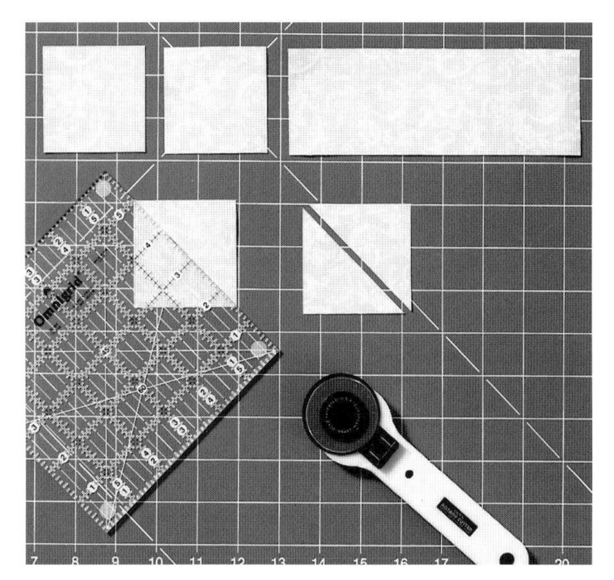

To compute the size square to cut, add ⅞" to the desired finished size of the short edge of the triangle. Cut a square this size and cut it once diagonally to yield two triangles. Once all seams are sewn, each triangle will be the desired finished size on the short edges.

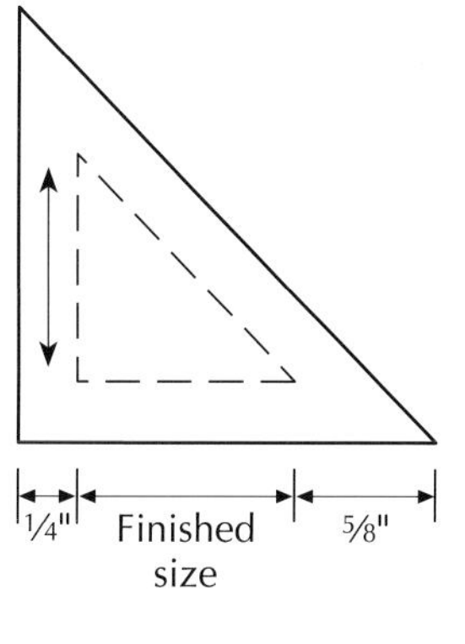

Quarter-Square Triangles

Make quarter-square triangles by cutting a square twice diagonally to yield four right triangles with the straight grain on the single long edge.

To compute the size square to cut, add 1¼" to the desired finished size of the long edge of the triangle. Cut a square this size and cut on both diagonals to yield four triangles. Once all seams are sewn, each triangle will be the desired finished size on the long edge.

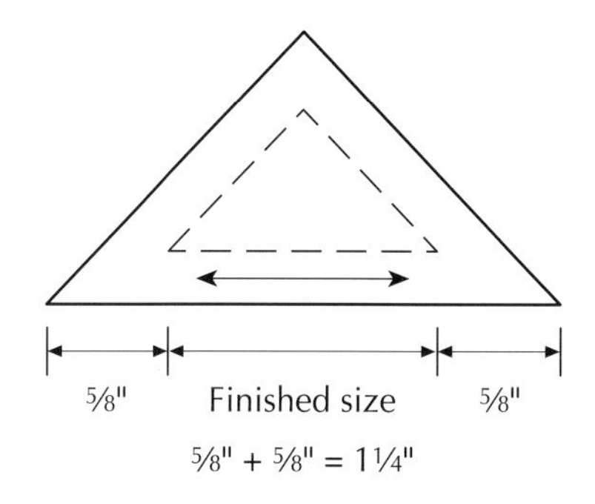

Side-Setting and Corner-Setting Triangles

When blocks are set diagonally, they often have triangles set in on the sides and corners to square off the piece. Because of grain-line needs, side triangles are cut from squares as quarter-square triangles. For the same reason, corner triangles are cut from squares as half-square triangles. It is best to cut these triangles slightly oversized and then trim the edges of the quilt or block to size later.

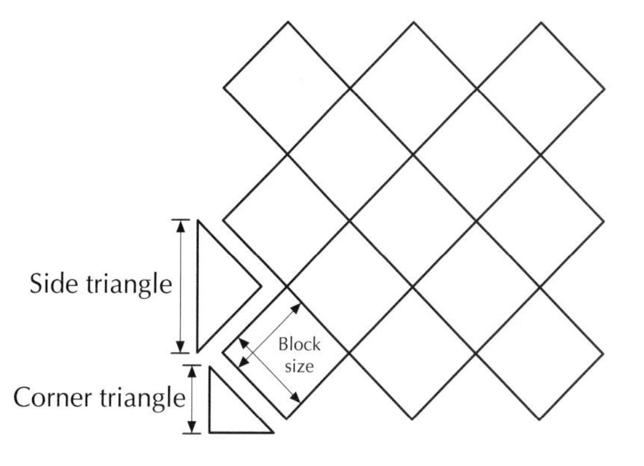

To cut the quarter-square side triangles for a quilt, you need to know the finished size of the triangle's long edge. To determine a slightly oversized measurement, multiply the quilt-block size by 1.5, then add 1¼" to this figure. Cut one square this size for every four side-setting triangles needed. Cut the squares into quarter-square triangles.

To cut the half-square corner triangles for a quilt, you need to know the measurement of the short edge of the triangle. To determine a slightly oversized measurement, multiply the quilt-block size by .75, then add 1" to this figure. Cut two squares this size and cut them into four half-square triangles.

Using Rotary-Cutting Aids

Up to this point, we have been using our rotary cutter and rulers to cut simple shapes, such as strips, squares, rectangles, and right triangles. These shapes all have easy numbers that can be added to the finished size to compute the cut sizes. But some shapes have no easy "magic" number to add to the finished size. Therefore, we need to use other methods or aids to rotary cut these shapes. I use one of two methods to achieve this goal: paper cutting guides, or a method I call mark, stack, and cut. Decide which method you prefer in each situation.

Paper Cutting Guides

A paper cutting guide is a cut-size template (finished-size pattern piece with a ¼"-wide seam allowance drawn around all sides) made from tracing paper. This template is then taped to the underside of a rotary ruler so you can: 1) cut appropriate strip widths, and 2) crosscut the strips into the desired shape.

The cutting guide must be translucent so it does not obstruct your view of the fabric underneath. It should also be very thin so as not to add additional thickness between the ruler and the fabric.

To cut strip widths, align and tape the straight-grain edge of the cutting guide to the edge of the ruler. If the shape has parallel edges as shown, cut fabric strips the width of the guide.

TIP

If the cutting guide is small, tape two cutting guides to the ruler to give yourself a longer distance with which to cut strips.

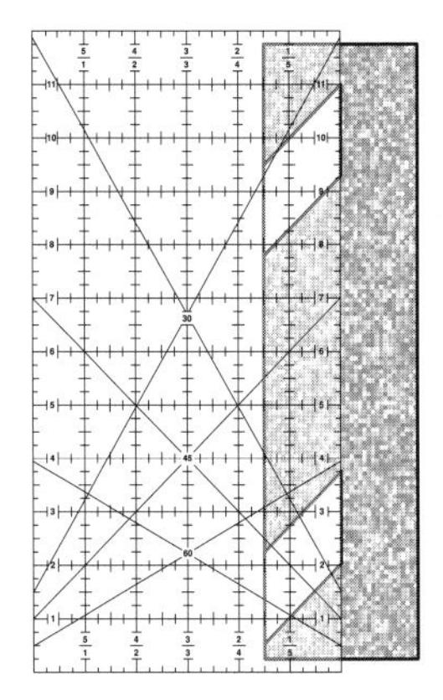

Use the guide to cut the individual units from the strip as shown below. Sometimes it takes two steps to cut each unit from a strip because of its odd shape or different angle. In some of these cases, it may be necessary to reposition the guide on the ruler to cut the units from the strip. Remember, the side of the cutting guide that you want to cut must be the side taped to the edge of the ruler.

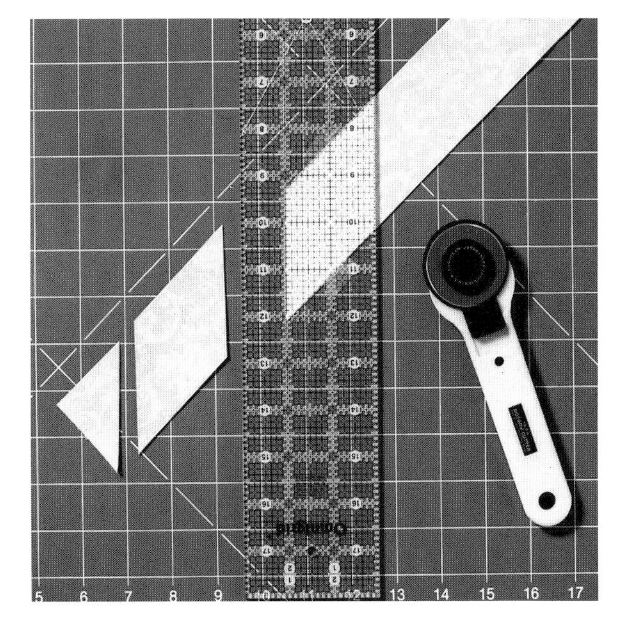

If the shape does not have parallel edges, as in the triangle shown, you will need to draw the unit on graph paper.

1. Draw the triangle on graph paper with seam allowances included. Then draw a line parallel to one of the unit's edges to form a more useful rectangular-strip cutting guide. It is difficult to use a single point to accurately cut a strip width.

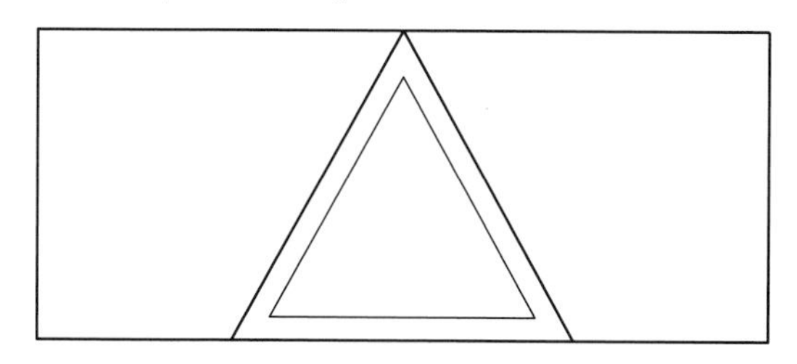

2. Cut out the rectangular guide and tape it to the ruler to cut the proper-size strip widths.
3. Once the strips are cut, remove the cutting guide from the ruler. Trim the excess used to make the strip cutting guide. Do not cut away the seam allowances.

4. To cut triangles, tape the new triangle guide to the ruler. Cut segments from the strips at the proper angle.

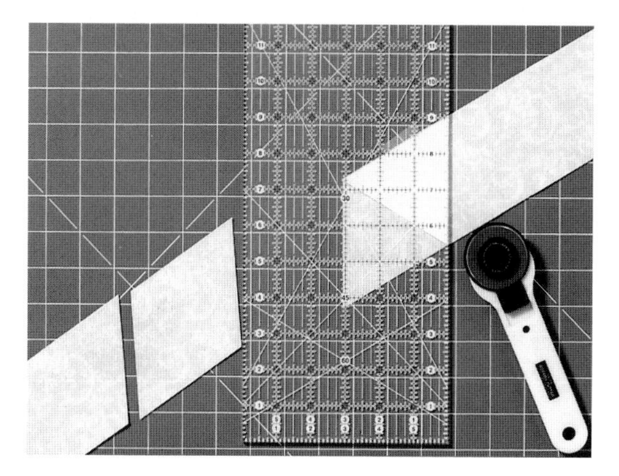

5. Turn the segments and cut them into the individual triangles, again using the cutting guide.

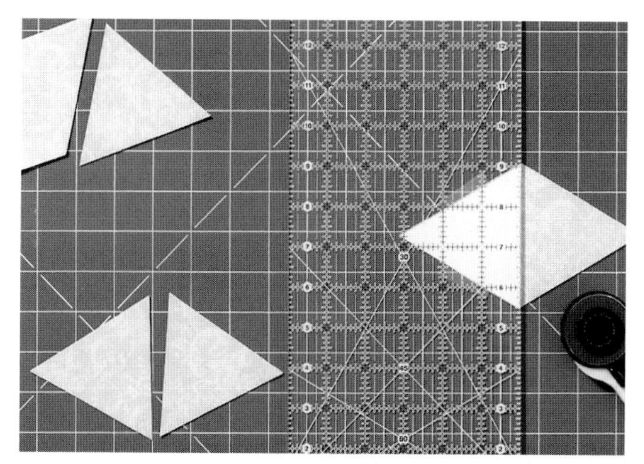

You can use paper cutting guides to rotary cut just about any straight-edged shape.

The Mark, Stack, and Cut Method

This method combines templates and rotary cutting. Some people prefer it to paper cutting guides. It is useful when you need only a few of one unit or are cutting one or two units each from many prints.

1. Draw a traditional plastic or cardboard template with seam allowances included.
2. Cut pieces of fabric slightly larger than the template shape. Cut 1 piece of fabric for every unit to be cut. If you want to cut several units from each print, you will need larger pieces of fabric.
3. Mark 1 fabric piece with the cut-size template and put it on top of a stack of 4 to 6 pieces.
4. Place your ruler on each marked line and carefully rotary cut the edges of the unit through the whole stack. Do not move the stack until all edges are cut, so the fabric layers don't shift.

Cutting Other Shapes

On-Point Squares

On-point squares are squares that are cut and set diagonally within a design, such as the Variable Star shown. It is important for the squares to measure evenly across the diagonal instead of along the sides. They must also have the straight grain on the edges of the square. There are two preferred methods to rotary cut these squares: 1) use a paper cutting guide or 2) cut strips and squares using the Bias Stripper ruler.

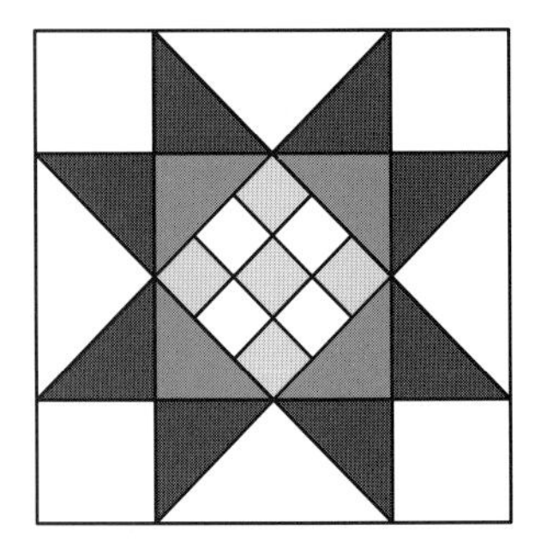

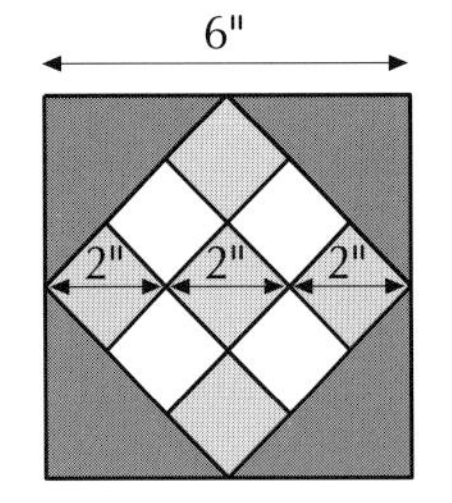

To fit, nine-patch squares should measure 2" across the diagonal.

To make a paper cutting guide, draw the desired square on-point and add seam allowances on all sides. Cut out the square to make a strip cutting guide. Refer to “Paper Cutting Guides” on page 14.

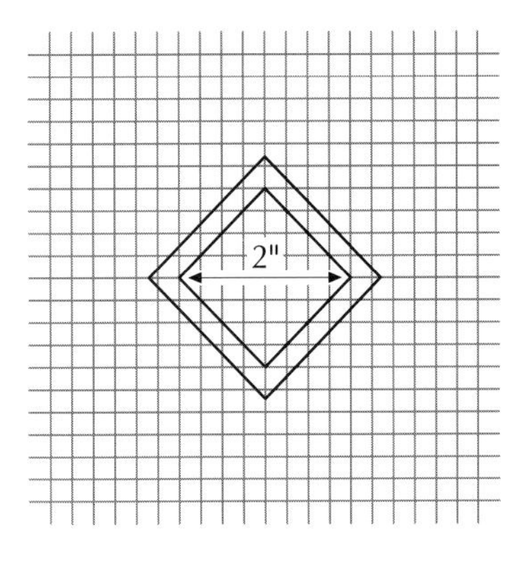

By far, the Bias Stripper is the easiest tool for the job. First determine the finished measurement of the square across its diagonal. Do not add anything to the finished size since seam allowances are already built into the unique measurements of the Bias Stripper. Then, using the Bias Stripper ruler, cut straight-grain strips at the ruler mark for the finished diagonal size. For example, for squares that measure 2" on the diagonal, cut the strips at the 2" mark.

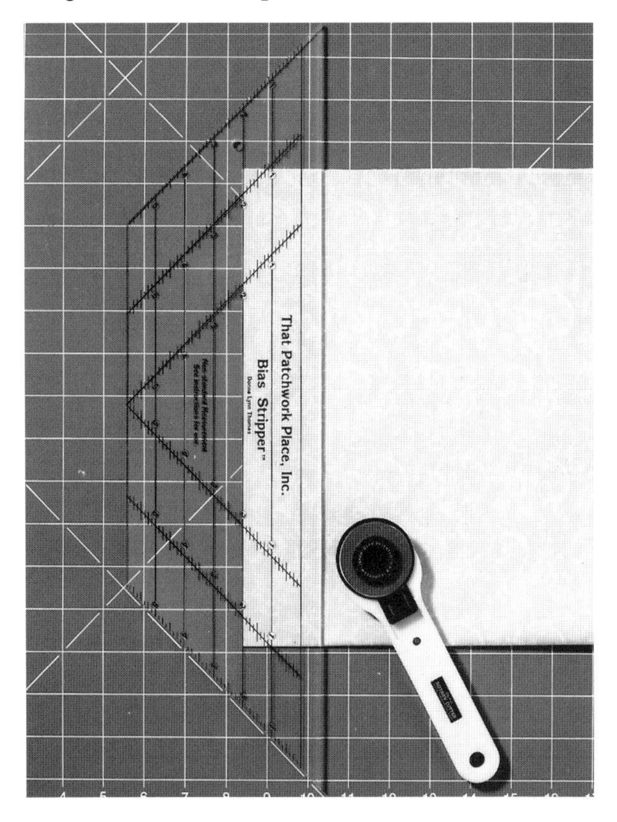

Then, crosscut the strips into squares at the same 2" mark.

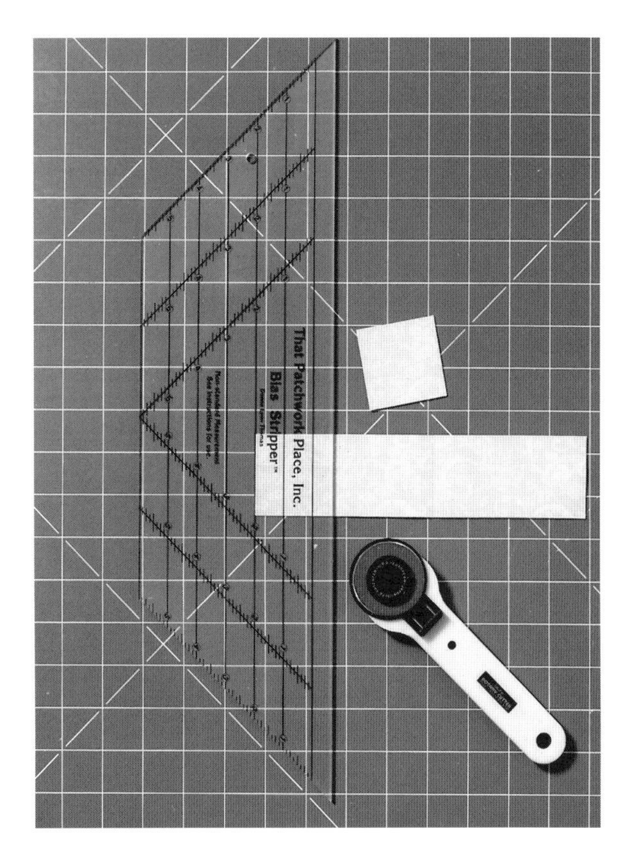

Equilateral Triangles

Equilateral triangles are triangles that measure the same length on each side, or leg. The angle at each corner is 60°. There are three ways to cut these triangles. You can use: 1) the 60° angle on your rotary ruler, 2) paper cutting guides, or 3) the mark, stack, and cut method.

Equilateral triangles can be rotary cut from straight-grain strips using just your ruler.

1. Add ¾" to the finished height of the triangle and cut straight-grain strip(s) this width.

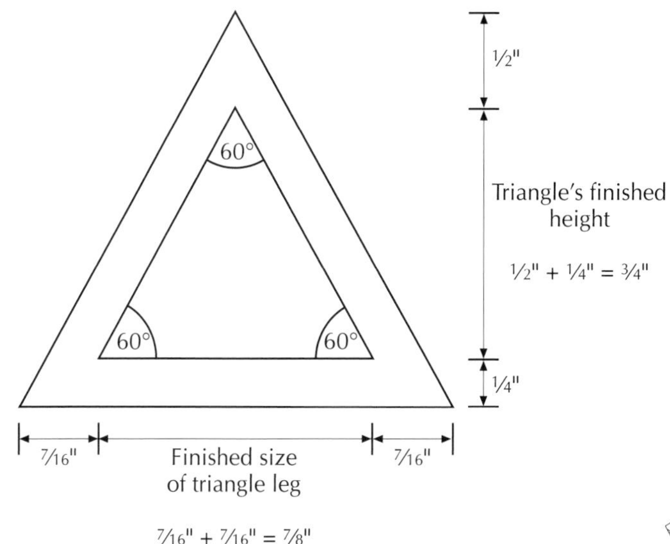

2. Add ⅞" to the finished size of the triangle leg. Mark this distance along the top edge of the strip for as many triangles as you need.

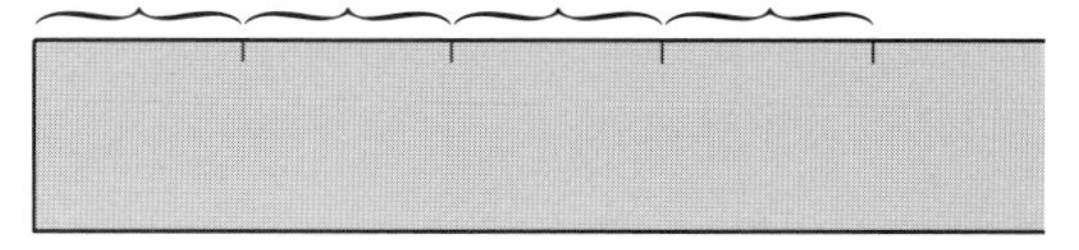

3. Align the 60° line of the ruler with the top edge of the strip so that the straight edge of the ruler intersects the first mark on the edge. Cut from raw edge to raw edge.

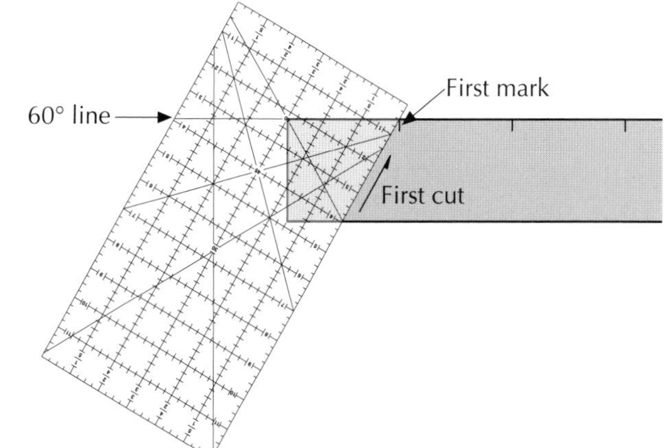

4. Rotate the ruler to align the 60° line with the edge you just cut so that the straight edge of the ruler intersects with the first mark on the edge. Cut from raw edge to raw edge. Continue cutting triangles across the strip, rotating the ruler as needed.

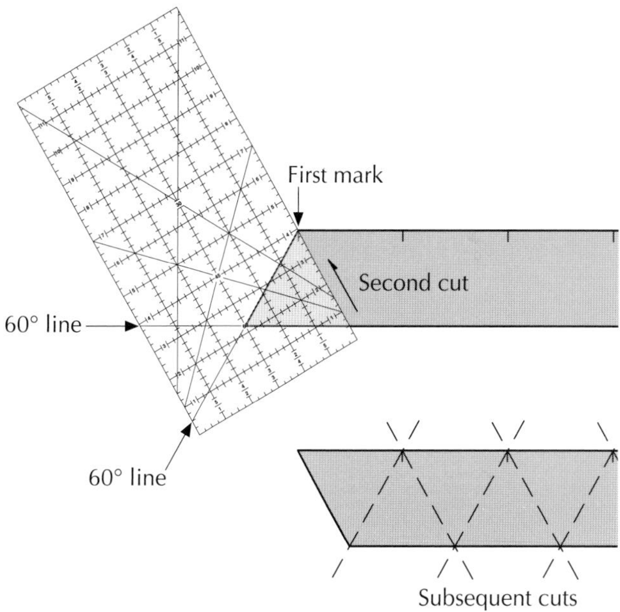

Refer to pages 14–15 for directions on how to use paper cutting guides or the mark, stack, and cut method.

Half-Rectangle Triangles

Make half-rectangles by cutting rectangles once diagonally. Please note that cutting on opposite diagonals results in mirror-image half-rectangles. Be careful to cut the diagonal in the direction that gives you the image you want. The best way to cut these triangles is to use a paper cutting guide.

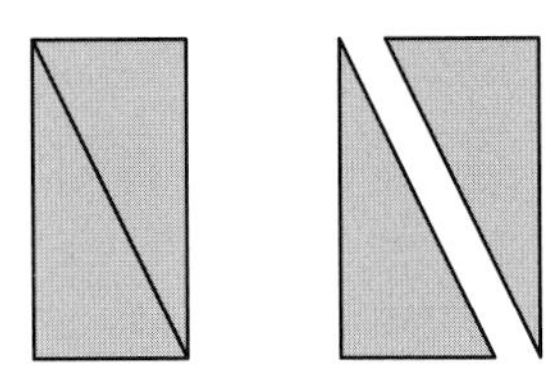

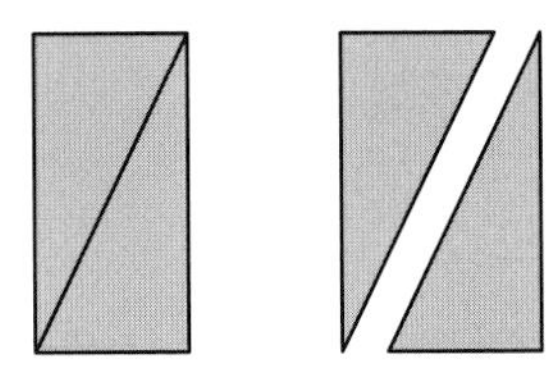

1. Draw the desired half-rectangle and add seam allowances on all sides. Add lines to turn it into a full rectangle. Use this shape as the strip cutting guide.

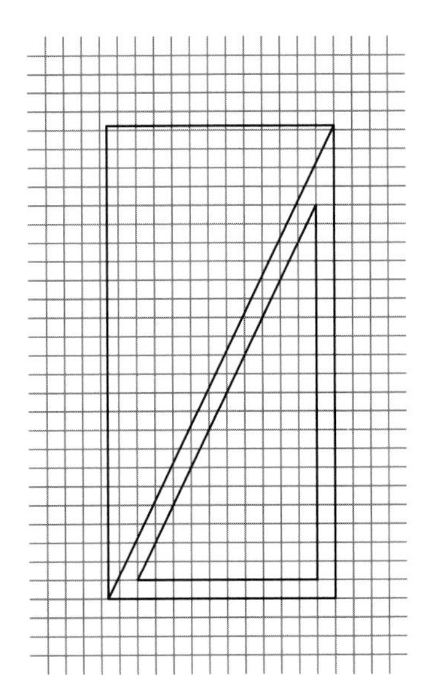

2. Cut out the rectangle guide and tape it to the ruler to cut the proper-size strip widths.

3. Use the same guide to crosscut the strips into rectangles. Cut the rectangles once diagonally to make half-rectangles.

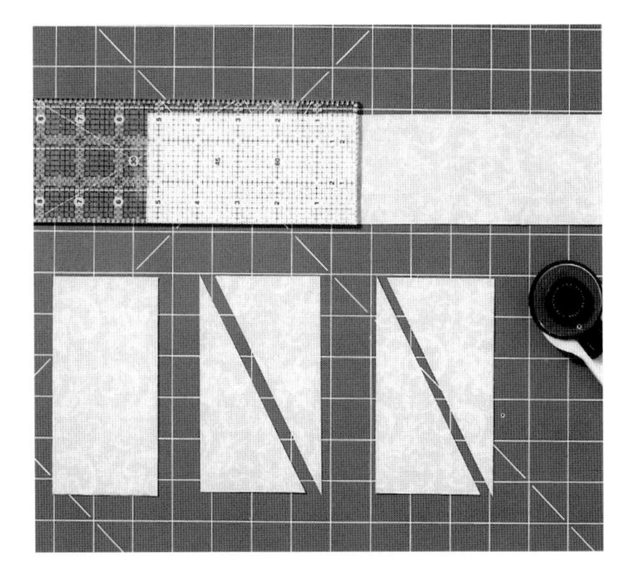

You can also use the mark, stack, and cut method on page 15 to cut half-rectangles. Be sure to mark the correct image on your fabric.

See *The Joy of Quilting*, by Joan Hanson and Mary Hickey, for information on a wonderful method for strip piecing half-rectangle units using Mary Hickey's BiRangle ruler.

Diamonds and Parallelograms

Parallelograms are four-sided shapes that have pairs of parallel sides. They are essentially squares tilted at an angle. A diamond is a type of parallelogram that measures the same length on all four sides. Quiltmaking uses other types of parallelograms with just the opposite sides measuring the same length, not all four sides.

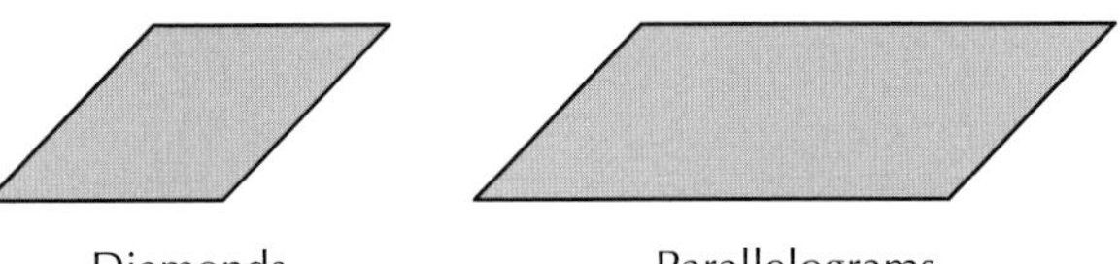

Diamonds come in three basic sizes: 30°, 45°, and 60°. The best method for cutting most diamonds and parallelograms is to use paper cutting guides. Refer to "Paper Cutting Guides" on page 14.

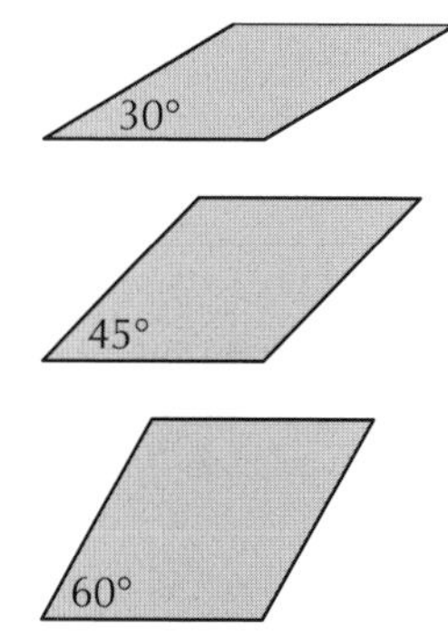

In addition to paper cutting guides, you can also cut 45° diamonds and parallelograms easily with the Bias Stripper ruler.

1. For diamonds, determine the desired finished size of the diamond along its edge. This measurement is the same size as the finished square or triangle that will be set in next to the diamond.

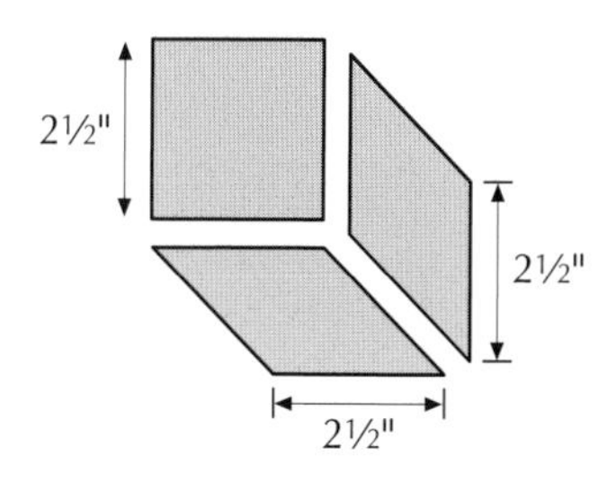

2. Cut straight-grain strips at the mark for the finished size, using the Bias Stripper. Using the example shown above, this would be the 2½" mark on the ruler. You do not need to add anything for seam allowances since this is already built into the unique measurements of the Bias Stripper.

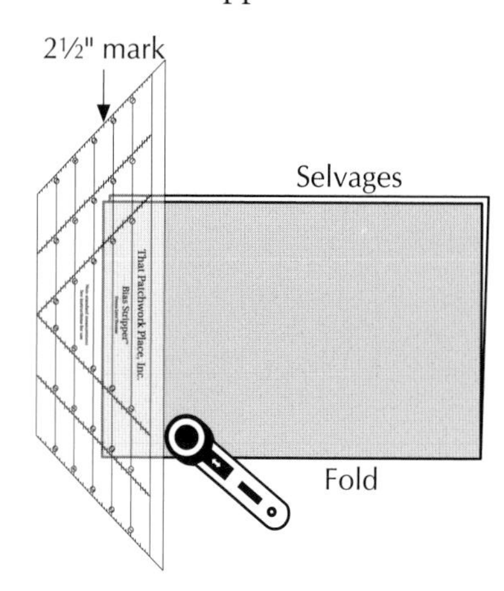

3. Unfold the strip. Align the Bias Stripper on the strip as shown and trim one end to a 45° angle.

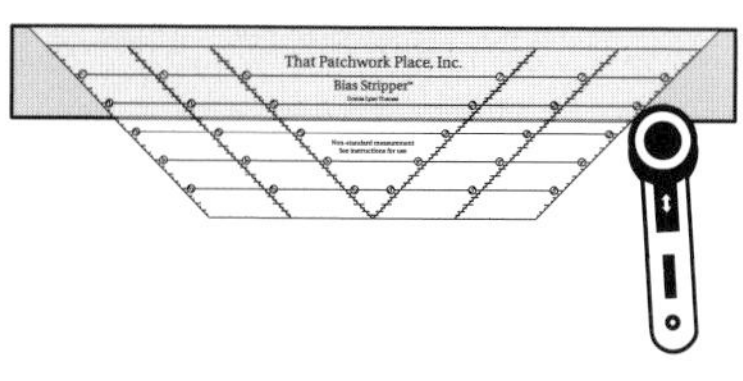

4. Turn the strip and cut diamonds from it at the same mark you used for cutting the strips. Align the ruler on both the cutting edge and the top raw edge of the strip when cutting. Retrim the end of the strip if you can't align the ruler on both edges.

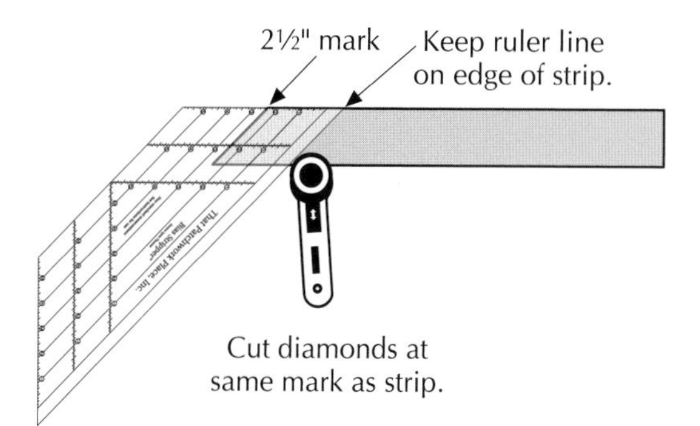

For parallelograms, the process is similar. First determine the finished size of one set of legs and cut straight-grain strips at this mark. Trim the strip to 45° as with diamonds. Determine the finished length of the second set of legs and cut the strips into parallelograms at this mark.

Trapezoids

A trapezoid is a unit that has two parallel sides and two nonparallel sides.

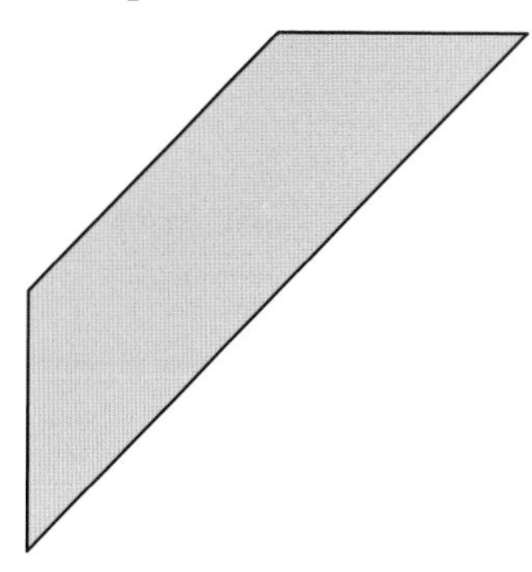

You can use a paper cutting guide to cut trapezoids from strips.

1. Draw the desired trapezoid and add seam allowances on all sides. Make a paper cutting guide from this drawing.

2. Cut fabric strips the width of the cutting guide.
3. Reposition the guide to cut trapezoids from the strips. Refer to "Paper Cutting Guides" on page 14.

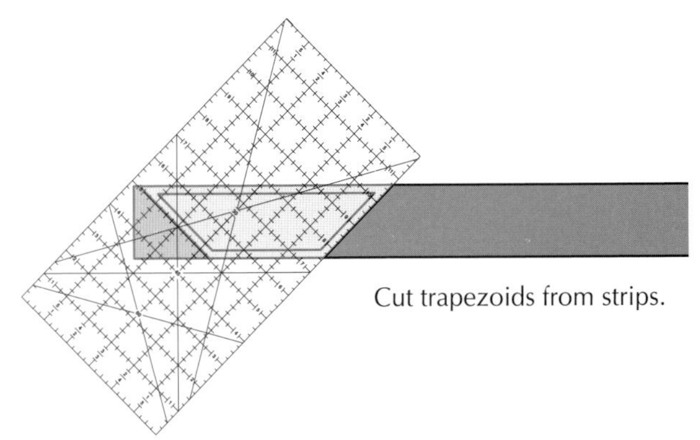
Cut trapezoids from strips.

Instead of strips, you can cut trapezoids from half-square triangles, using either a paper cutting guide or the Bias Stripper ruler.

1. Draw the finished-size trapezoid. Do not add seam allowances. Extend the 2 nonparallel sides until they meet, forming a finished-size triangle.
2. Add ⅞" to the finished size of the triangle and cut a square this size.
3. Cut the square once diagonally to make 2 half-square triangles.

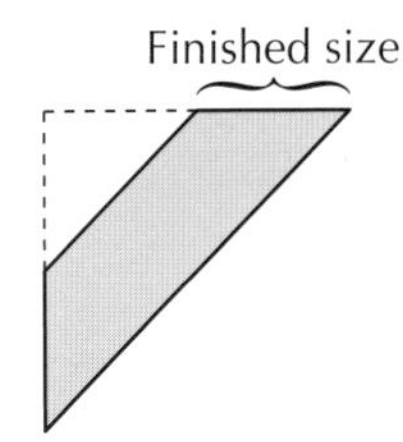

To use a paper cutting guide, tape the short, parallel edge of the guide to the edge of the ruler and lay it on a triangle. Trim the small corner that sticks out beyond the ruler edge. This is an excellent method to use when you need only a few trapezoids.

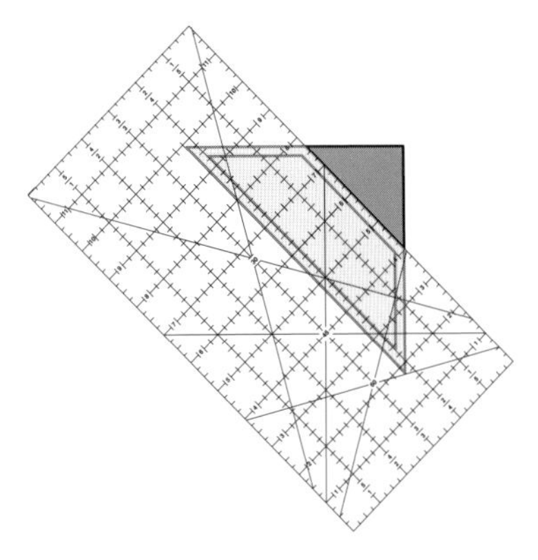

To use a Bias Stripper ruler, determine the finished size of the short, nonparallel edge of the trapezoid, and place the Bias Stripper ruler on the long edge of the triangle at this mark. For example, if the short, nonparallel edge is 2", place the 2" mark of the ruler on the edge of the triangle. Trim the excess corner.

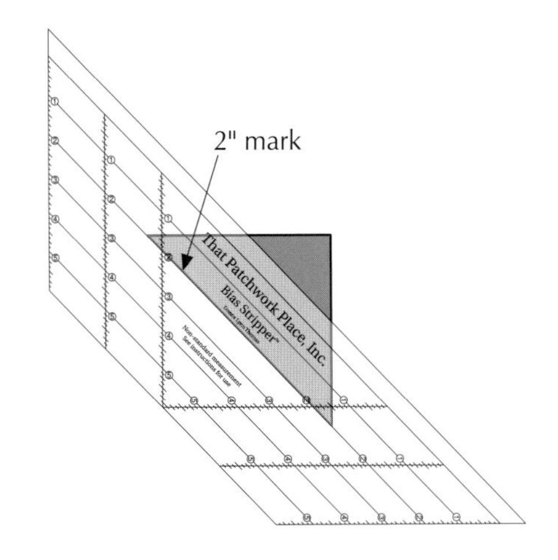

Octagons

Octagons are best cut from squares of fabric using a paper cutting guide.

1. Draw the octagon and square it off. Add seam allowances to the square and cut a fabric square this size.

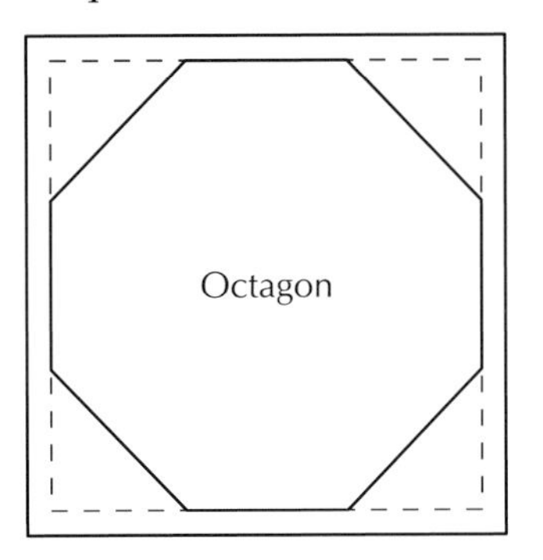

2. Add seam allowances to the drawing of the original octagon. Make a paper cutting guide from this shape.

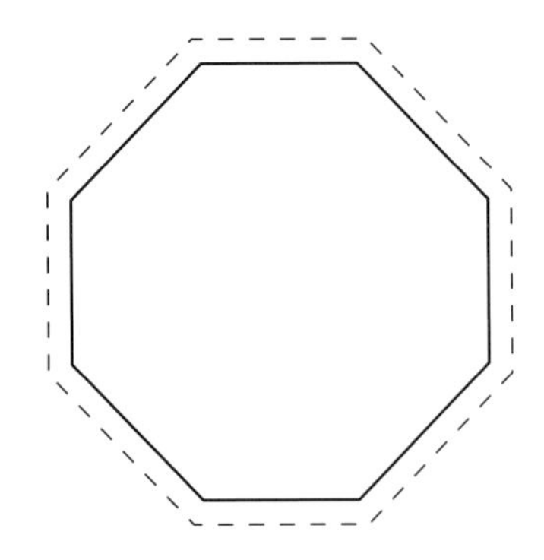

3. Tape the corner of the octagon cutting guide to the ruler and use it to remove the corners from the square of fabric, creating an octagon.

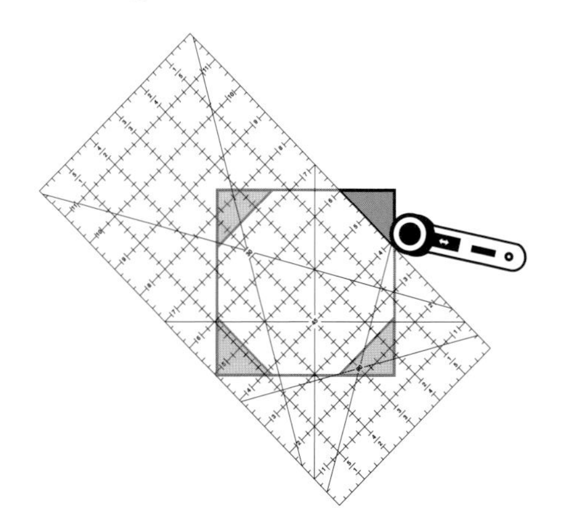

Nubbing

Nubbing points removes excess fabric that extends beyond the ¼"-wide seam allowances, making the edges of pieces easier to match for more accurate stitching.

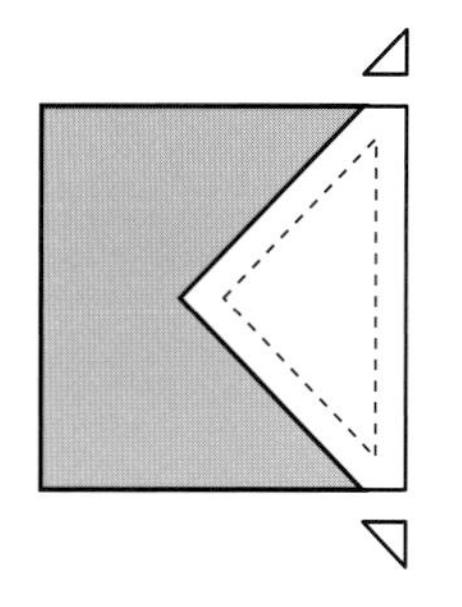

For shapes other than half- and quarter-square triangles, begin by making translucent-paper templates with sewing lines marked and seam allowances included. Lay the paper pieces right sides together as if they will be sewn, aligning the sewing lines. Pin to secure the edges and trim the excess paper beyond the seam allowances. In the example shown, nub only the long point of the half-rectangle triangle. Use these trimmed paper templates as nubbing guides to trim the cut fabric pieces to size.

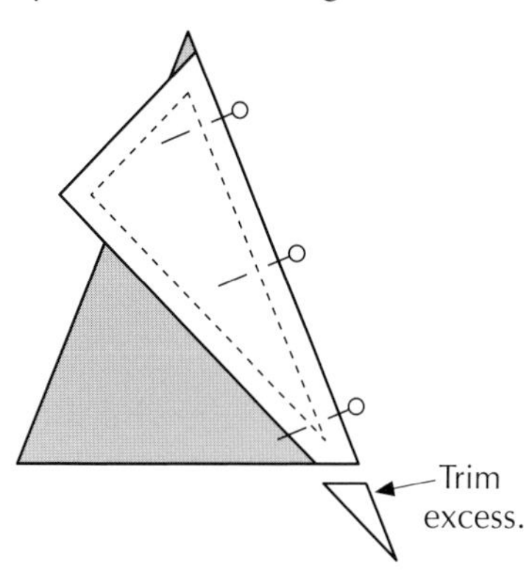

Nub half-square and quarter-square triangles, using the Bias Square ruler or other rotary ruler.

To nub half-square triangles, add ½" to the finished size of the triangle's short side. For example, add ½" to a 1½" finished-size triangle to get 2". Place the Bias Square ruler on the corner of the triangle at the 2" mark as shown. Cut off tips that extend beyond the ruler's edges.

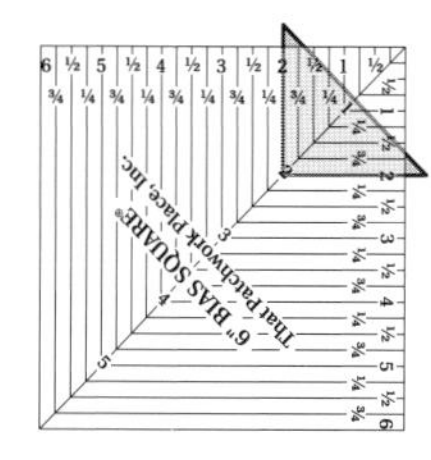

To nub quarter-square triangles, add ¼" to half of the finished size of the triangle's long side. For example, on a triangle with a 5" long side, add ¼" to 2½" (half of 5") to get 2¾". Fold the triangle in half. Place the Bias Square ruler on the fold of the triangle at the 2¾" mark as shown. Cut off the tips that extend beyond the ruler on the right side only. *Do not cut the tips at the top of the fold!*

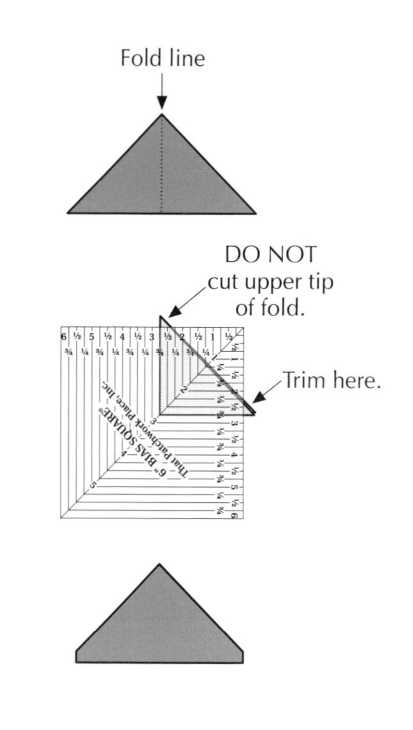

Modifying Units

There are many new and easy ways to create some of our frequently used pieced units. Bias strip piecing is one method to create many of these and it is dealt with completely in my book *Stripples Strikes Again!* Besides bias strip piecing, there are other accurate and easy ways to modify existing units to create new pieced units.

These techniques generally fall into two categories: 1) sewing squares onto squares, rectangles, and triangles and 2) sewing triangles onto triangles. The basic concept is to sew smaller squares or triangles onto the corners of larger units to create a new unit.

Read through the different applications of this technique presented here. Once you understand the concept behind it, I'm sure you'll run across other situations where you can apply it.

Sewing Squares onto Squares, Rectangles, and Triangles

The basic process to modify shapes is the same for most units. Let's use a simple square with a corner triangle as an example to demonstrate the process.

You could assemble this unit by first cutting a square, cutting off a corner, and then sewing an individual triangle to the trimmed corner. Unfortunately, this method tends to be fraught with problems. It is much easier and less frustrating to modify a larger square into the pieced unit.

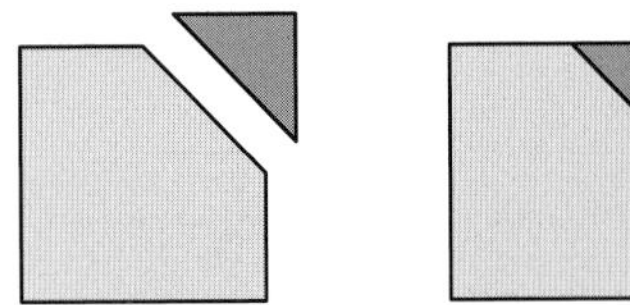

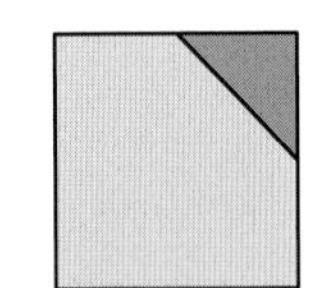

1. Draw the desired unit to determine the finished sizes. In the example below, the pieced unit will finish to 4". Cut a square ½" larger than the finished size (4" + ½"= 4½").
2. Determine the finished size of the triangle edge where it falls on the side of the pieced square. Add ½" to this measurement and cut a square this size (2½").

Finished size of triangle is 2".

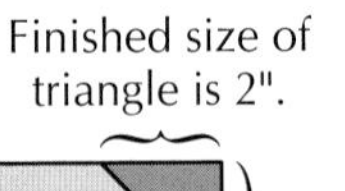

Finished size of square is 4".

3. Draw a diagonal line on the wrong side of the small square from corner to corner. Place the fabric on top of very fine sandpaper if you find the fabric shifts as you mark. Place the small square on the corner of the large square with right sides together as shown. Pin. Sew on the marked line.

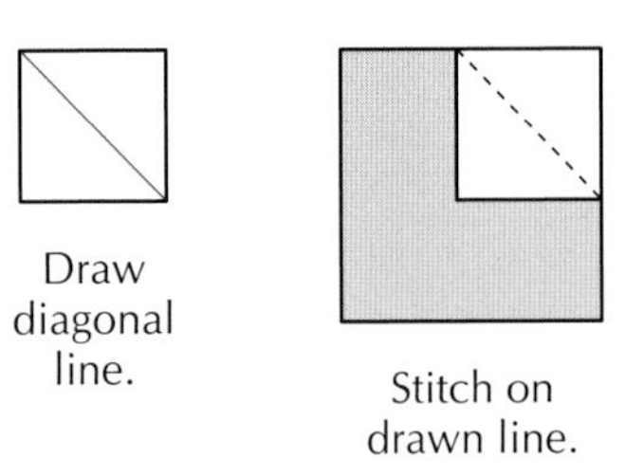

Draw diagonal line.

Stitch on drawn line.

4. Press the small square back over the corner and check it for accuracy. It should lie over the corner of the large square. If not, adjust the seam until it does. Some people find it more accurate to sew just barely to the right of the marked line. Experiment to see what works best for you.

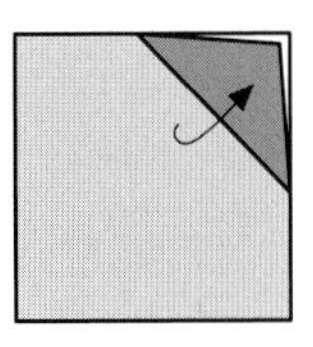

Fold square over corner.

You now have three layers of fabric in the corner. You have several choices when it comes to trimming the excess behind the corner triangle.

- Do not trim anything. The corner of the original large square can be used as the sewing edge when sewing the unit into the block. This is okay if you don't mind the bulk and plan to machine quilt through the layers.
- Trim the center layer only. You still have the corner of the large square as a reference when sewing, but less bulk remains.
- Trim both layers from behind the top triangle. This is best if you are planning to hand quilt and you are sure your triangle is accurate as a sewing edge.

This is the basic process used to modify squares, rectangles, and triangles with smaller squares. You will be amazed at the variety of pieced units you can create with this simple technique.

Square-within-a-square: Many of our pieced blocks contain a pieced center that consists of a square set on-point within another square. You can make this unit by sewing small squares to all four corners of a large square.

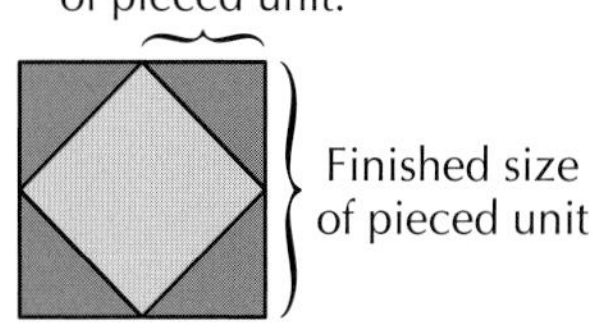

Cut a large square ½" larger than the finished size of the pieced unit. Cut it from the fabric that will be used for the center square, set on-point.

For a square-within-a-square, the corner triangles are always equal to half the size of the finished unit. As above, add ½" to this measurement and cut squares this size for the corners.

Follow the basic process on page 22 to sew the small squares to the corners of the large square. Be sure to sew, press, and trim one corner square completely before sewing a square to the adjacent corner.

NOTE: You can sew squares to any number of corners to create other types of pieced units.

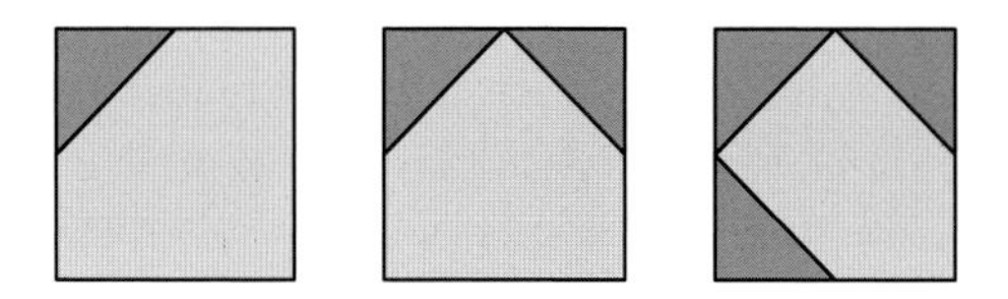

Snowball: This unit is similar to a square-within-a-square except that the points of the triangles do not meet on the sides of the square. Use the basic process on page 22 to sew the small squares onto the larger square.

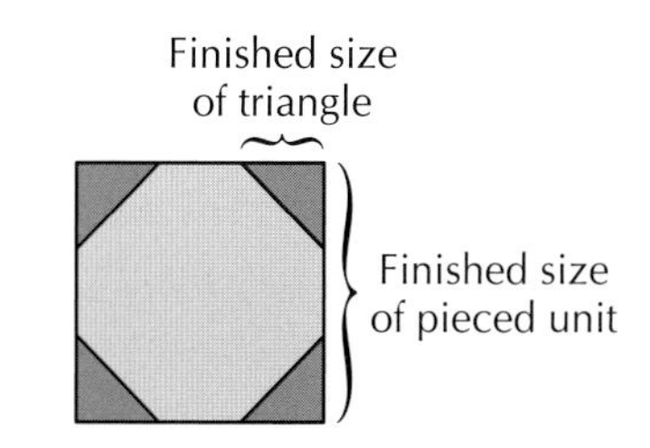

Pieced square: Traditionally composed of two half-square triangles, this unit can be assembled in a variety of ways. One of the simplest methods makes two pieced triangles at a time.

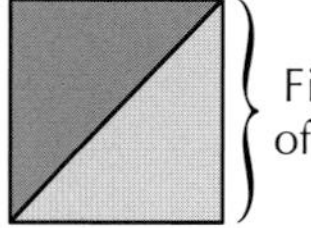

1. Add ⅞" to the desired finished size of the triangle (this is also the finished size of the pieced square). Cut 1 square this size from each of the 2 prints used in the pieced square.
2. Draw a diagonal line on the wrong side of 1 square and place it right sides together with the other square. Pin. Sew ¼" from both sides of the diagonal line.

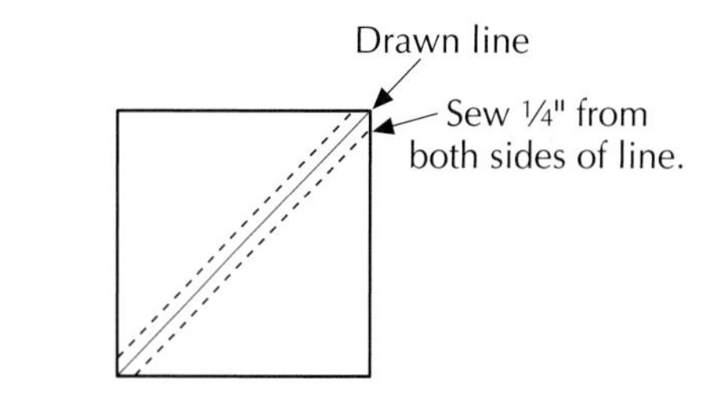

3. Cut on the diagonal line. Press the units to form 2 pieced squares.

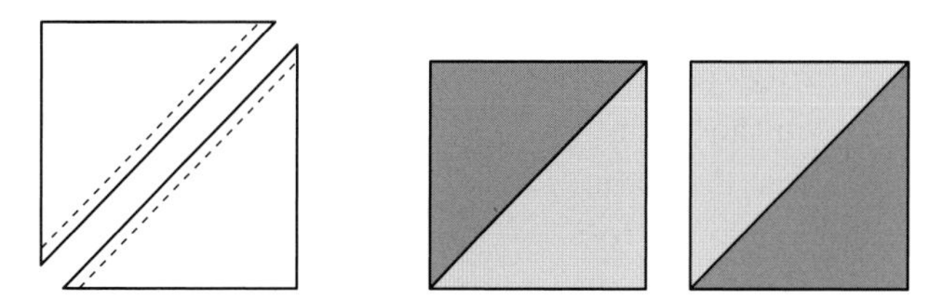

TIP

If your pieced squares are consistently off-size or distorted, add 1" to the finished size of the triangle instead of ⅞". Cut squares this size and follow the instructions on page 23 to make two slightly oversized pieced squares. Cut each of the pieced squares back to the proper cut size (finished size + ½"), using the Bias Square ruler. Although an additional step is required, the result will be perfectly sized pieced squares every time.

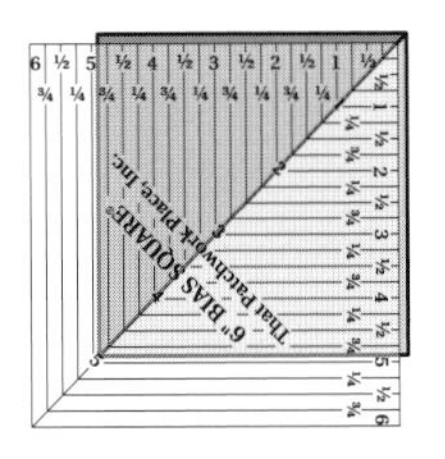

Place diagonal line of ruler on seam line. Trim first two sides.

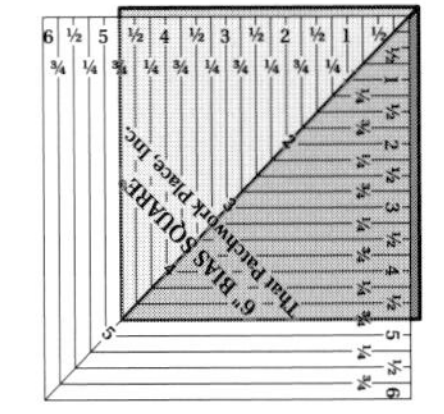

Align desired measurement on previously cut edge, and diagonal line on seam. Trim remaining sides.

Striped square: This type of pieced square is routinely found in patchwork blocks. An easier way to make this unit than the traditional method of assembling three individual pieces is to modify a pieced square. The following method makes two striped squares at a time. Two steps are involved in the process: 1) making the pieced squares and 2) sewing the smaller squares onto the corner.

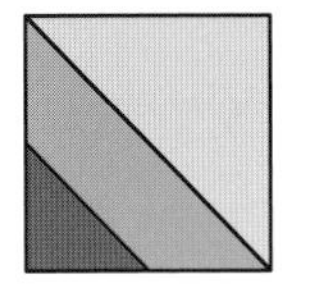

Striped square

1. Draw the striped square. Add ⅞" to the finished size of the striped square and cut 2 squares this size from each of the 2 prints that will form the center seam. Using these squares, make a pair of pieced squares as described on page 23.

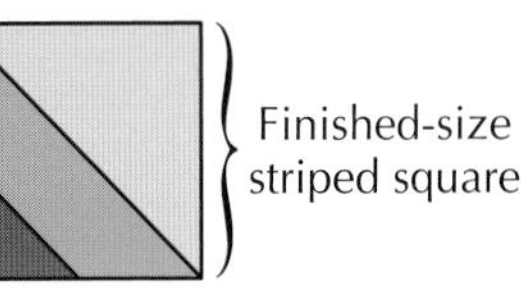

2. Cut a small square that is ½" larger than the finished size of the small corner triangle on the striped square. Cut 1 for each pieced square. Sew a small square to the corner of each pieced square to make 2 striped squares.

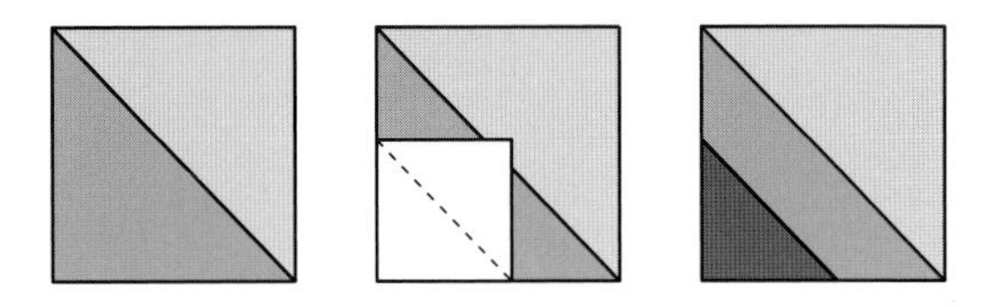

Pieced rectangle: Rectangles have right-angle corners like squares. Many pieced blocks use rectangles that have triangles sewn onto their corners. Use the basic process on page 22, sewing small squares onto the corners, to modify them.

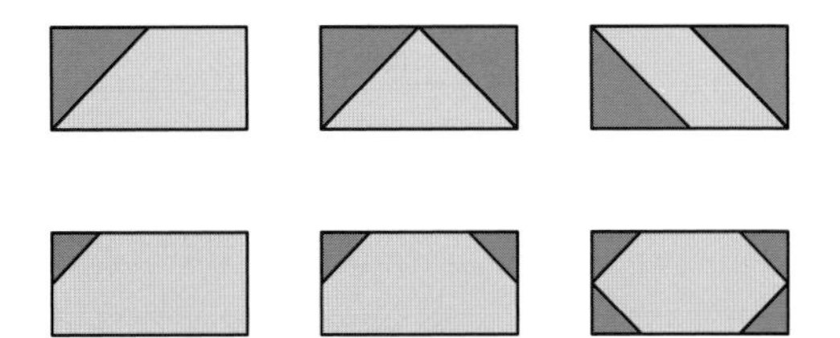

The assembly process is the same as with squares. Be sure to make an accurate drawing of the unit to determine the finished size of the triangles so you can cut the proper-size corner squares.

Pieced triangle: In the same way that you sew small squares onto the corners of large squares, you can sew a small square onto the corner of a large triangle to make a pieced triangle.

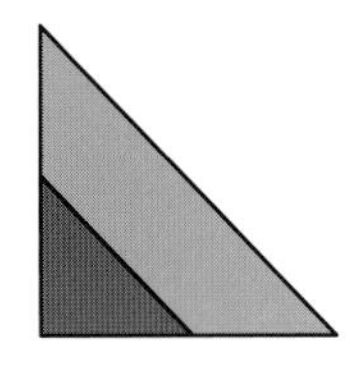

Pieced triangle

1. Draw the pieced triangle. Add ⅞" to the finished size of the pieced unit and cut a square this size. Cut the square once diagonally to yield 2 half-square triangles.

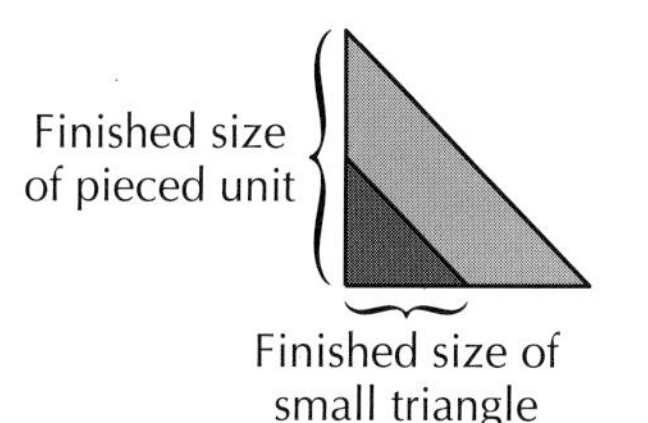

2. Cut a small square that is ½" larger than the finished size of the small triangle. Sew a small square onto the corner of the large triangle to make a pieced triangle.

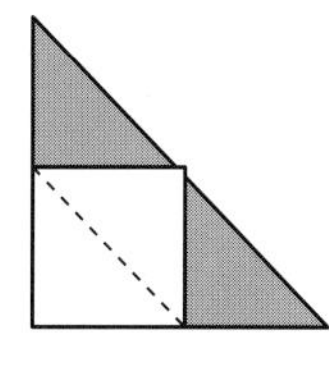

Sewing Triangles onto Triangles

You can also create other types of pieced triangles by sewing smaller triangles onto the corners of larger plain or pieced triangles. The advantages are the same in that you eliminate a lot of frustration and fussy piecing while increasing the accuracy of the finished product.

The basic process for sewing smaller triangles onto larger triangles is similar to sewing smaller squares onto corners of larger squares. You are simply working with a different shape. Since the units are based on half-square triangles, you can make two units at a time. A simple pieced triangle can be used as an example.

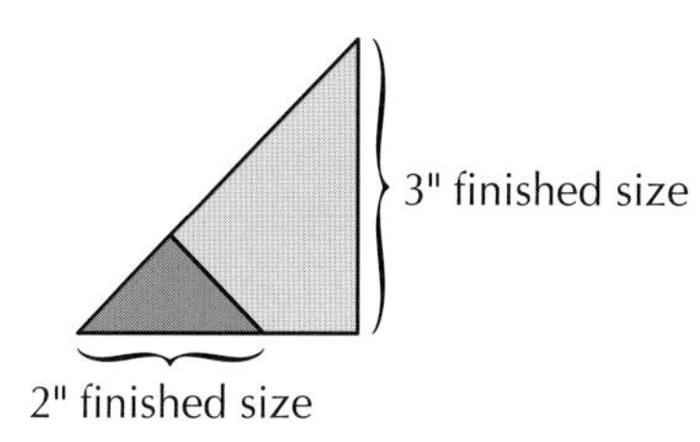

1. Draw the desired unit to determine the finished sizes. In the example shown, the finished size of the entire pieced triangle is 3". The finished size of the small triangle, where it lies on the short edge of the large triangle, is 2".
2. For the large triangle, add ⅞" to its finished size and cut a square this size. Cut the square once diagonally to make 2 large half-square triangles.
3. For the small triangle, add ⅞" to its finished size and cut a square this size. Draw a diagonal line on the wrong side of the square from corner to corner. Cut the square in half on the other diagonal to yield 2 marked half-square triangles.

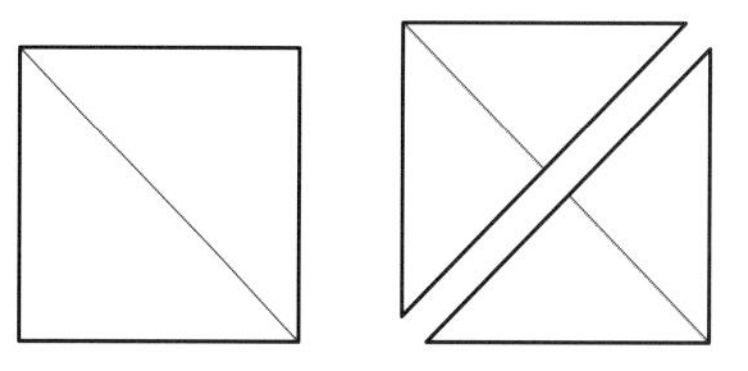

4. Place the small triangle on the corner of the large triangle, right sides together. Pin. Sew on the marked line. Be sure to sew the small triangle on the correct corner to obtain the desired pieced triangle. Notice the difference between the 2 pieced triangles shown.

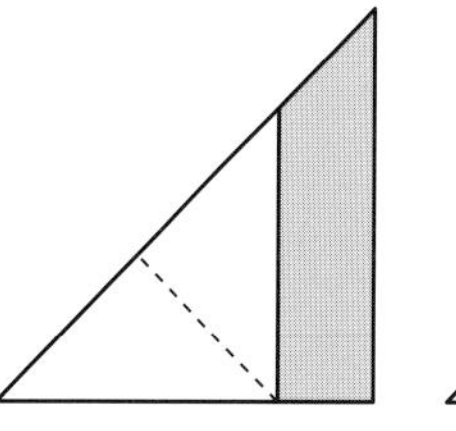
Stitch on line.

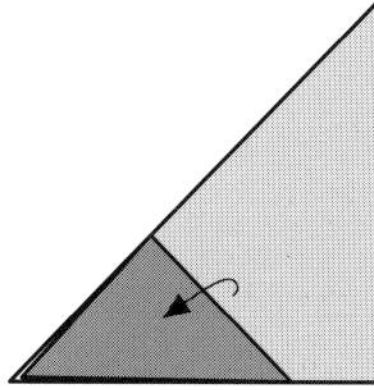
Fold triangle over corner.

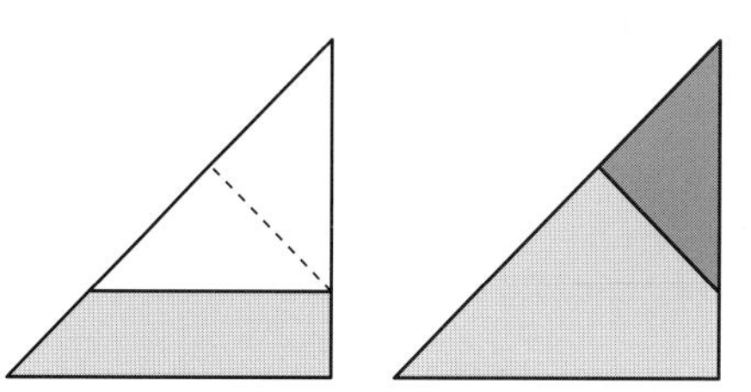

5. Press the triangle back over itself on the corner and check it for accuracy. It should lie over the corner of the large triangle. If not, adjust the seam until it does. You have the same 3 trimming choices as you did with sewing squares.

NOTE: Sew a small triangle to the opposite corner of the large triangle to create another type of pieced triangle.

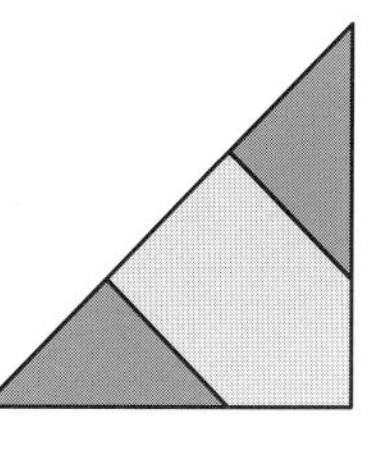

Striped triangle: Just as you made striped squares, you can make striped triangles by sewing a small triangle on the corner of a larger pieced triangle. The following method makes four striped triangles at a time. There are two steps to the process: 1) making the pieced triangle and 2) sewing the small triangle onto its corner.

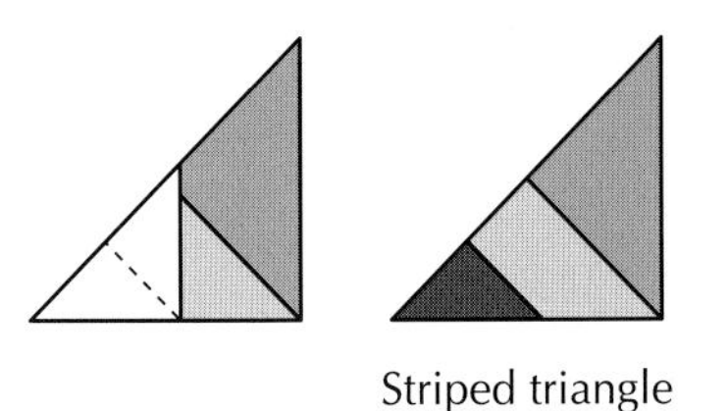

Striped triangle

1. Draw the striped triangle to determine finished sizes. Add 1¼" to the finished size of the striped triangle and cut 1 square this size from each of the 2 prints that will make up the pieced triangle. Cut each square twice diagonally to yield 4 quarter-square triangles.

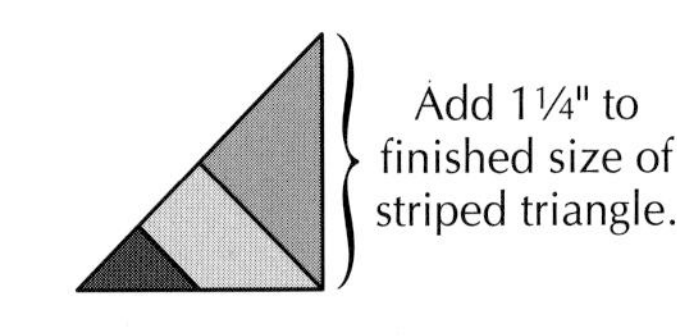

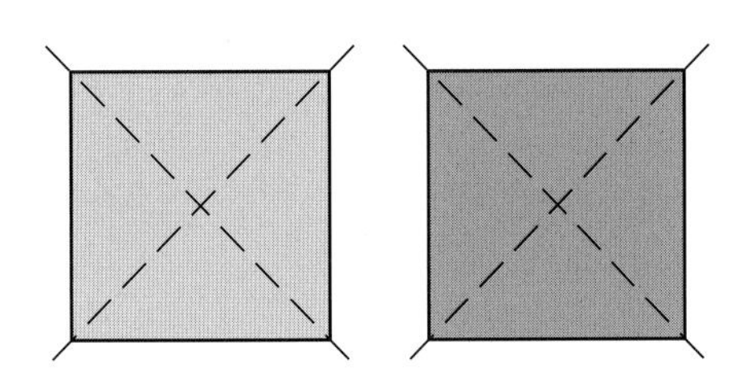

2. Sew the quarter-square triangles together to make 4 pieced triangles.

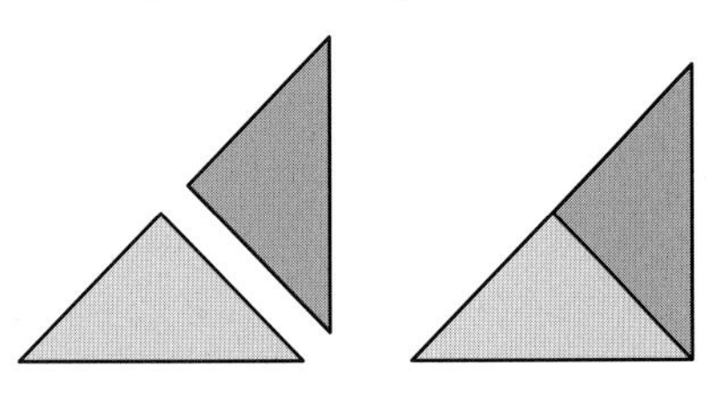

Pieced triangle

3. Add ⅞" to the finished size of the small triangle, where it lies on the short edge of the pieced unit. Cut 2 squares that size and draw a diagonal line from corner to corner. Cut the squares in half on the other diagonal to yield 4 marked smaller triangles. Sew and press a small triangle onto the corner of each pieced triangle.

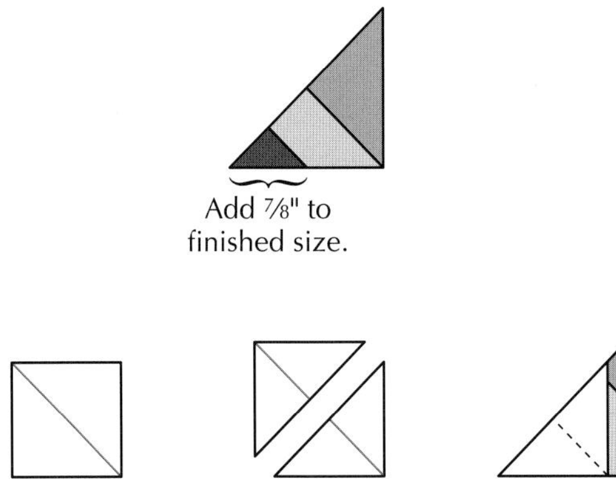

Using Strip-Piecing Techniques

Straight-Grain Strip Piecing

Straight-grain strip piecing is a simple concept. The goal of this technique is to eliminate the need to cut and sew individual pieces, thereby speeding up the assembly process and improving accuracy.

A good example of straight-grain strip piecing is shown in the simple Nine Patch block. Normally, you would cut and sew nine individual squares together to make the block. But if you break the block into its three rows, you can see that rows 1 and 3 are identical and row 2 is composed of the same shapes and colors but in a different arrangement.

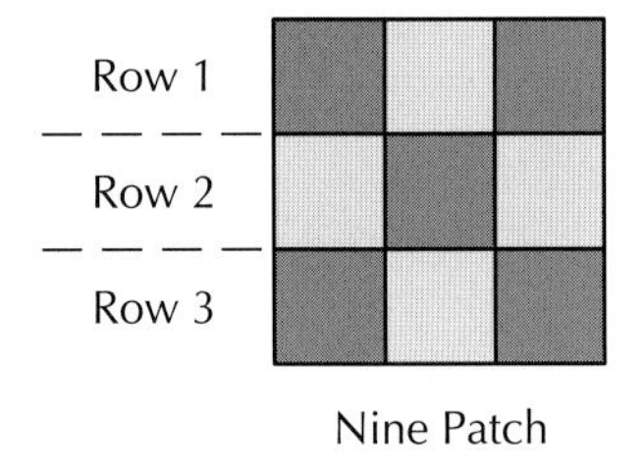

Nine Patch

Instead of squares, you can cut straight-grain strips of fabric and sew them together side by side, then cut them into segments that look like the rows in the Nine Patch block. It is a simple matter to sew the segments together to complete the block.

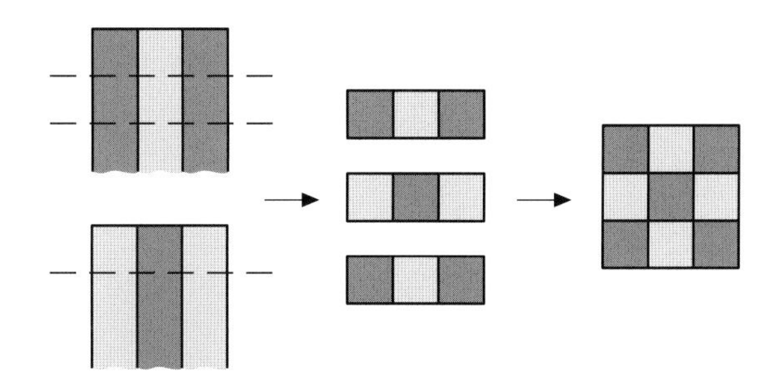

Assembling Straight-Grain Strip Units

Straight-grain strip units are made of two or more straight-grain strips sewn together side by side.

1. Cut strips perpendicular to the selvage, across the fabric width, in the width given for the quilt you are making.
2. Remove the selvages from all strips before using them in a strip unit.
3. Cut all full-length (42") strips in half to 21" before working with them unless instructed otherwise. Full-length strip units tend to curve from sewing on the cross grain, making them more difficult to cut accurately into segments. Half-length strip units tend to curve less and are easier to handle.
4. Press seams as you go. It's difficult to press all the seams in a strip unit after it's completely assembled.

Pressing Straight-Grain Strip Units

There are two primary reasons for pressing seams in a particular direction: 1) so that seams rest against each other (butt) where they meet at intersections and 2) so your completed block or quilt will lie flat and smooth. Pressing toward the darker fabric is a luxury you can consider after these two conditions are met.

To see this concept at work, let's look at the Nine Patch block. To successfully butt seams and evenly distribute the seam allowances, you must press the strip units in one of two ways. The first is to press the seams in rows 1 and 3 in one direction and the seams in row 2 in the opposite direction. The second way is to press the seams toward the dark strips since they alternate in this block. Always plan your seam directions so all seams butt at intersections.

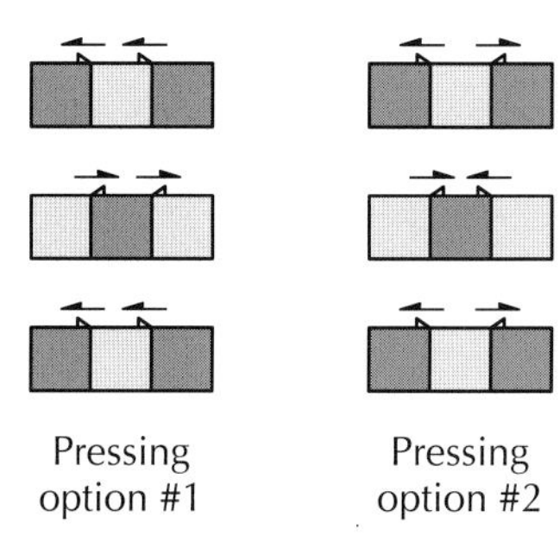

Cutting Segments

1. Make a clean cut along the short edge of the strip unit. This is your cutting edge. It is important to make all cuts at right angles to the strip-unit seams, or you will have cockeyed squares and rectangles. Be sure to line up the horizontal ruler lines on the strip seams to make a right-angle cut.

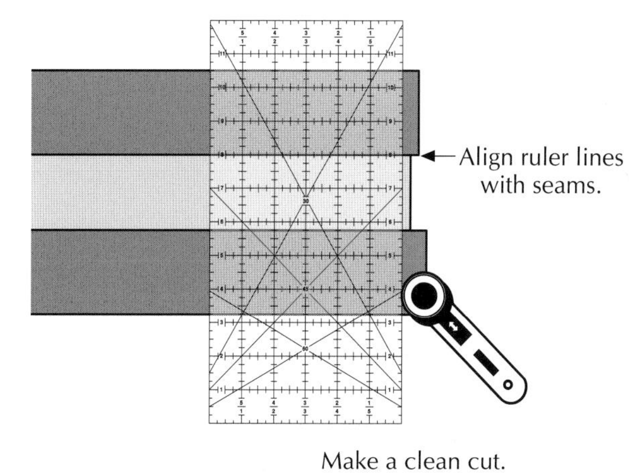

Make a clean cut.

2. Turn the strip around and position the ruler the required distance from the cutting edge for the segments. Don't forget to line up the horizontal ruler lines on the strip seams to make a right-angle cut. Working from left to right, cut segments across the strip unit.

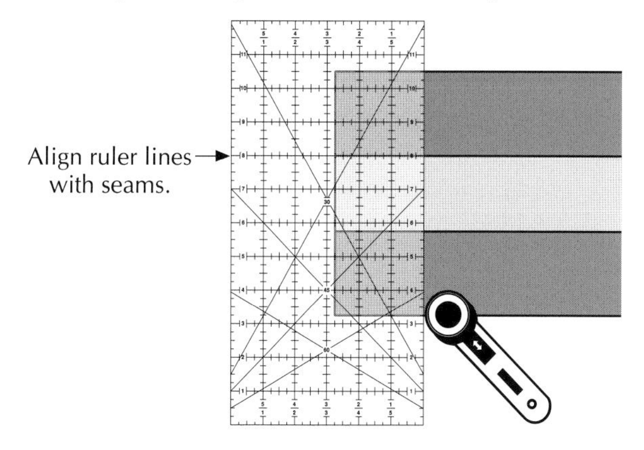

3. If, at any time, the short edge and interior seams no longer form a right angle, make a new, clean cut on the short edge. It is normal to make this adjustment periodically due to the miniscule amount of ruler slippage that occurs with each cut. The more carefully you cut, the less frequently you'll have to trim the short edge.

Basic Bias Strip Piecing

Basic bias strip piecing is another method used to make bias squares (pieced squares). A bias square is composed of two right triangles joined on the long edges. This unit is used extensively in both traditional and contemporary quilt patterns. You learned how to cut single triangles earlier in the book and also how to make a pair of these units from squares of fabric. Now we'll explore a method to cut these units from already sewn and pressed strip units.

Bias strip piecing is similar to straight-grain strip piecing except that the strips are cut on the bias grain instead of on the straight grain. Why cut strips on the bias? A look at the grain lines on a bias square provides the answer. The straight grain should lie on the edges of the square, and the bias along the seam.

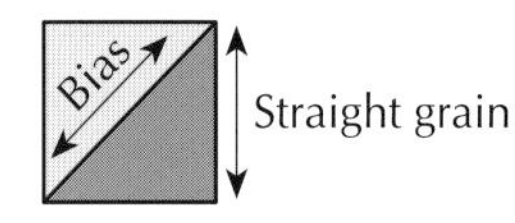

If you cut a square from straight-grain strip units, the outer edges of the squares will have bias edges. But if you cut a square from bias strip units, the outer edges of the square will have straight-grain edges.

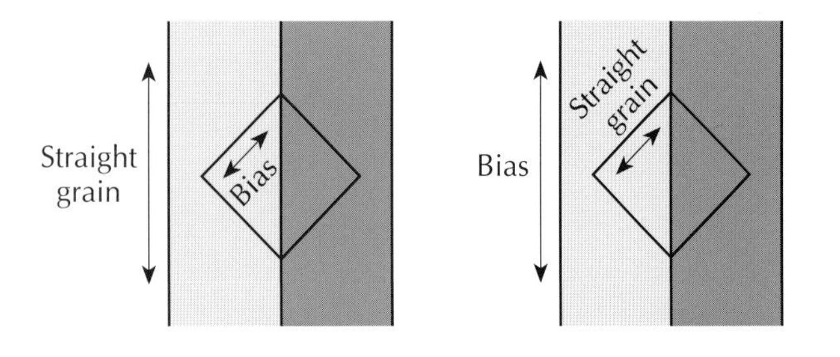

Cutting and Sewing Bias Strips

So how do you cut strips with bias edges? If you take a large square cut on the straight grain and cut it on the diagonal, you create two triangles with the bias on their long edges. You can cut bias strips from these long edges.

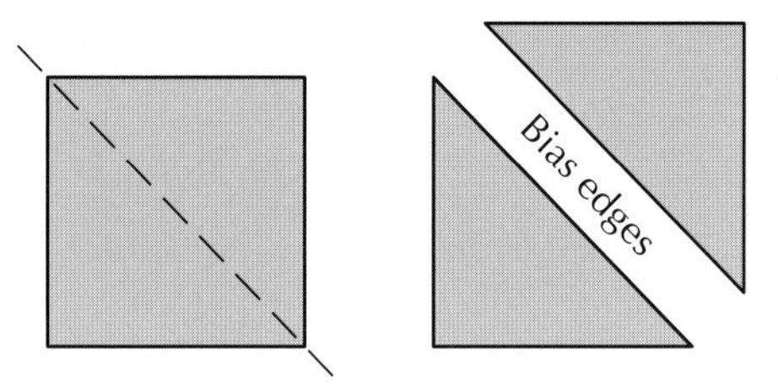

1. Cut a large square from each of the 2 prints you want in your bias square. Layer the 2 squares of fabric with right sides together.
2. Cut the layered squares once diagonally. Cut bias strips from the long edges of both sets of triangles.

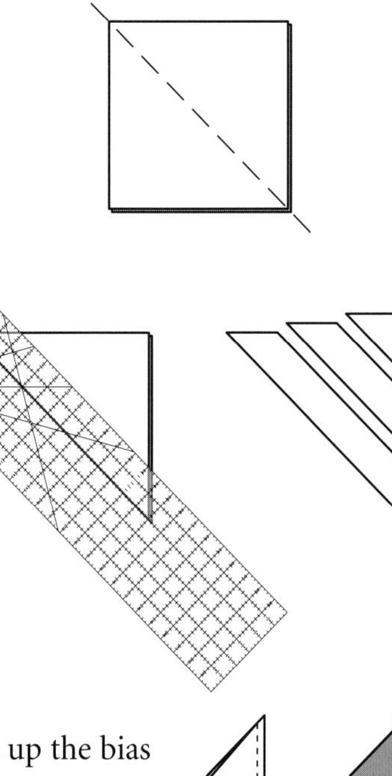

3. Pick up the bias strips in pairs as you cut them, and sew them together along their long bias edges. Press.

continued on page 37

GALLERY

DAYLILIES PARTY AT NIGHT by Deb Rose, 1998, Lansing, Kansas, 64" x 64". This stunning Lily quilt captures the full range and glory of the venerable daylily. Bright colors set on a deep blue background create a stunning contrast. The cheerful border print, which is the source for color choices in the rest of the quilt, ties all the fabrics together beautifully. Directions begin on page 56.

RHUMBA STAR by Donna Lynn Thomas, 1998, Doylestown, Pennsylvania, 36" x 36". This sparkling line of dancing stars was the result of a classroom mishap. A student, Ann Schoen, accidentally laid out the units for another star with the star points going in the wrong direction. We thought the stars looked like they were line dancing and so, with a few more changes, Rhumba Star was born. Quilted by Kari Lane. *Directions begin on page 39.*

MELODY TO THE RHUMBA BEAT by Gabriel Pursell, 1998, Havertown, Pennsylvania, 40" x 40". Gabriel's quilt takes the musical theme one step further with the pieced border. The blue squares are actually musical notes that travel around the border, ending with the peach notes—the end of the musical phrase. I am always so impressed with the creative talent of quilters!

STARS IN STOWE by Ann Woodward, 1998, Collegeville, Pennsylvania, 42" x 42". Ann chose reproduction prints to create a lovely, appealing quilt. Notice how she alternated dark and light fabrics in the blocks to create a checkerboard look. She quilted it while on her yearly summer trip to Stowe, Vermont.

CHRISTMAS SPLENDOR by Dee Glenn, 1998, Moorpark, California, 66" x 66". This complex-looking design is actually just sixteen Optical Illusion blocks set together without sashing, forming a wonderful overall circular design. Dee's choice of Christmas colors and prints makes this spectacular holiday quilt sparkle. *Directions begin on page 52.*

DELTA ZETA GIRLS by Linda Kittle, 1998, Leavenworth, Kansas, 56¾" x 56¾". Linda chose the bright, light colors of her college sorority to make this elegant quilt. Besides the optical illusion created by the blocks, Linda added a clever illusion of her own. By placing light-colored prairie points between the blocks and the first border, Linda makes the light-colored portion of the block design appear to extend into the borders.

WALRUS IN THE MIRROR by Kari Lane, 1998, Lawson, Missouri, 56" x 56". An interesting print with walruses in it was the inspiration for this rich-looking quilt. Kari's diverse collection of batiks adds the strong contrast needed to make this pattern work.

SEVEN SISTERS by Donna Lynn Thomas, 1998, Doylestown, Pennsylvania, 30" x 26¼". Traditionally made with diamonds, the Seven Sisters design represents the original seven southern states that seceded. This variation has 150 triangles instead of diamonds, and the use of at least seventy prints adds sparkle to the design. It's an easy way to use lots of leftover prints! Directions begin on page 46.

RIBBONS by Kari Lane, 1998, Lawson, Missouri, 30" x 26¼". Intending to lay out the pieces into a baby block design, Kari saw more ribbons in the end result. If you look at the quilt a minute or two, you'll see both images move forward and recede alternately. The result is a delightfully interesting quilt.

SISTERS #2 by Kari Lane, 1998, Lawson, Missouri, 30" x 26¼". In this sample, Kari alternated blue and fuchsia star points on a light background for a slightly different effect. Again, lots of prints make the piece dazzling.

NEVER TOO LATE by Robin Chambers, 1998, Media, Pennsylvania, 88" x 88". This bold and beautiful quilt is full of life and movement. The colors for the blocks were all drawn from the lovely floral print used in the big triangles and border. Not only do the blue triangles in the block spin, but the rectangles at the four corners of the block form a windmill when the blocks are set together without sashings. *Directions begin on page 48.*

COSMIC WHIRLIGIG by Ursula Reikes, 1998, Ivins, Utah, 63¾" x 63¾". Ursula's delightful and happy quilt would be perfect for a child's room. The primary colors dance and whirl across the surface of the quilt in mischievous red and blue swirls. The charming striped border pulls it all together perfectly. Quilted by Janice Nelson.

CHRISTMAS LILIES by Donna Lynn Thomas, 1998, Doylestown, Pennsylvania, 36" x 36". In this variation of "Daylilies Party at Night," traditional red and green is used to create a holiday wall hanging. A pointed border made from pairs of bias squares frames the quilt like a picture. The holly-leaf quilting complements the theme perfectly. Quilted by Ann Woodward.

LEGACY by Donna Lynn Thomas, 1997, Peachtree City, Georgia, 34" x 34". The traditional Album block in this quilt is rendered in beautiful old-time florals and Liberty of London prints. The center of each block is the perfect place to record signatures, inspirations, and dates. Quilted by Judy Keller. *Directions begin on page 42.*

Basic Bias Strip Piecing – *continued from page 28*

Cutting Bias Squares

Use a Bias Square ruler to cut bias squares from strip units. The Bias Square ruler has a diagonal line and ⅛" markings that meet in the center to form squares at ⅛" increments.

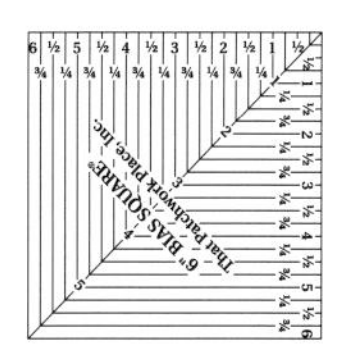

1. Begin at the lower end of the bias strip unit and position the diagonal line of the Bias Square ruler on the seam. The numbers should be at the top. Make sure the desired dimensions are just inside the raw edges at the bottom of the strip unit.
2. Cut the top 2 edges of the bias square from raw edge to raw edge.

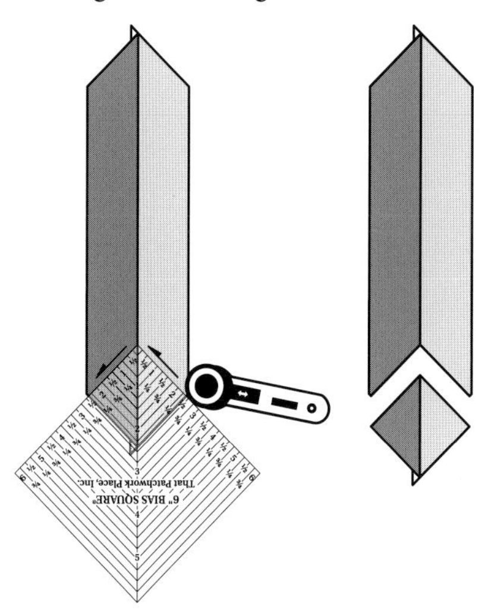

3. Turn the bias square and trim the remaining 2 raw edges to the proper size by aligning the desired markings on the Bias Square ruler with the clean-cut edges.

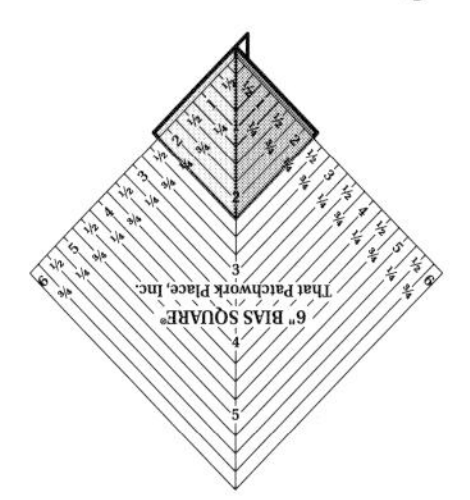

4. Move up the strip unit, cutting bias squares as you go. There will be leftover triangles created along the edge of the strip.

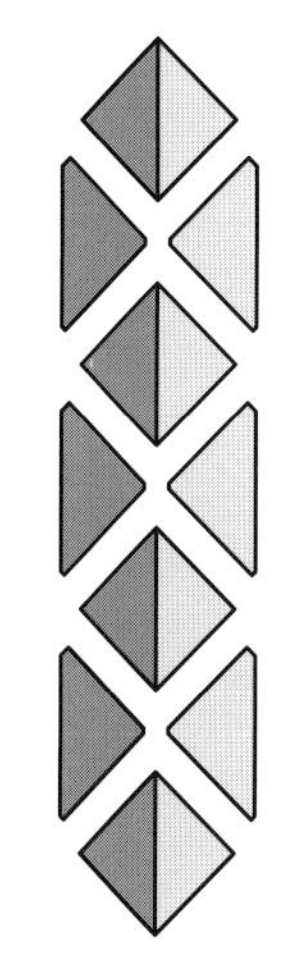

5. If you do not need to use these edge triangles, you can reduce their numbers by sewing single pairs of bias strips into multiple bias strip units. Fewer outside edges mean fewer edge triangles.

   Cut bias squares from multiple bias strip units in the same fashion as single pairs of strips. But instead of moving up the strip unit, cut the lowest points first across the bottom of the strip unit. Once the bottom row is cut, begin cutting the lowest points on the next row up as shown below.

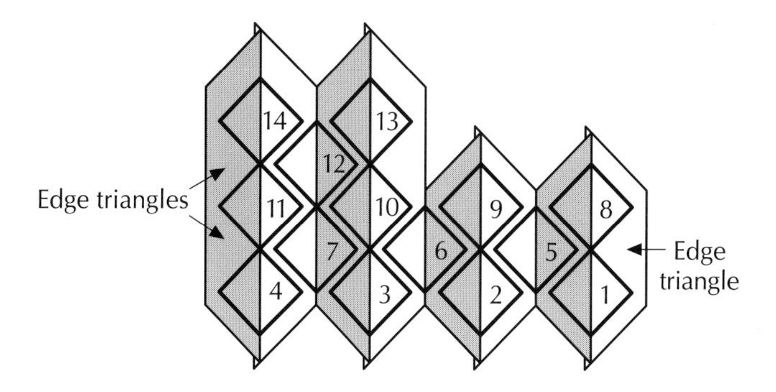

If you have a Bias Stripper Ruler, determine the strip width for any bias square by adding ¾" to the desired finished size of the bias square. Cut bias strips at this mark using the Bias Stripper.

If you don't have a Bias Stripper, you must use a mathematical formula to determine the proper strip widths to cut with a regular ruler. Multiply the desired finished size of the bias square by 1.414. Divide this figure in half, round it up to the nearest ⅛", and add 1". Cut bias strips this width, using a regular ruler.

The Bias Stripper has the math built into its design and eliminates the need to know the formula. Please see my book *Stripples Strikes Again!* for more extensive information on bias strip piecing and its many uses beyond bias squares.

Use the chart below to determine bias strip width and the resulting bias-square yield for two commonly used sizes of bias squares.

Finished Size	Cut Size	Bias Strip Width	Yield from 9" square	Yield from 12" square	Yield from 15" square
2"	2½"	2½"	13	21	35
3"	3½"	3⅛"	7	13	18

Please note that the yield figures are based on sewing multiple bias strips and are close approximations only. Actual yields may vary from person to person.

The Quilts

Now it's time for the fun stuff! There are a few things for you to remember before you get started.

- Each quilt project begins with a list of techniques used in that particular quilt. Be sure to go back and review the information about each technique before beginning the quilt.
- Take the time to make sure you have an accurate seam allowance (see page 5). This is probably the most important thing you can do when it comes to quiltmaking.
- Fabric requirements for the quilts are based on fabric that measures at least 42" wide from selvage to selvage after washing. If your fabric measures less than this, you may need more fabric.
- All cutting measurements include ¼"-wide seam allowances.
- All cutting dimensions are for a regular ruler unless you are instructed to use a Bias Stripper ruler or the cutting guides provided.
- Press all seam allowances in the direction of the arrows unless otherwise instructed.
- Please read the general information at the back of the book, starting on page 60, to learn how to assemble and finish your quilt.

If you work through all the patterns in *Shortcuts*, you will have mastered quite a few rotary-cutting techniques and, hopefully, had a lot of fun in the process. Enjoy!

Rhumba Star

Color photo on page 30

Finished Quilt Size: 36" x 36"
Finished Block Size: 8" x 8"

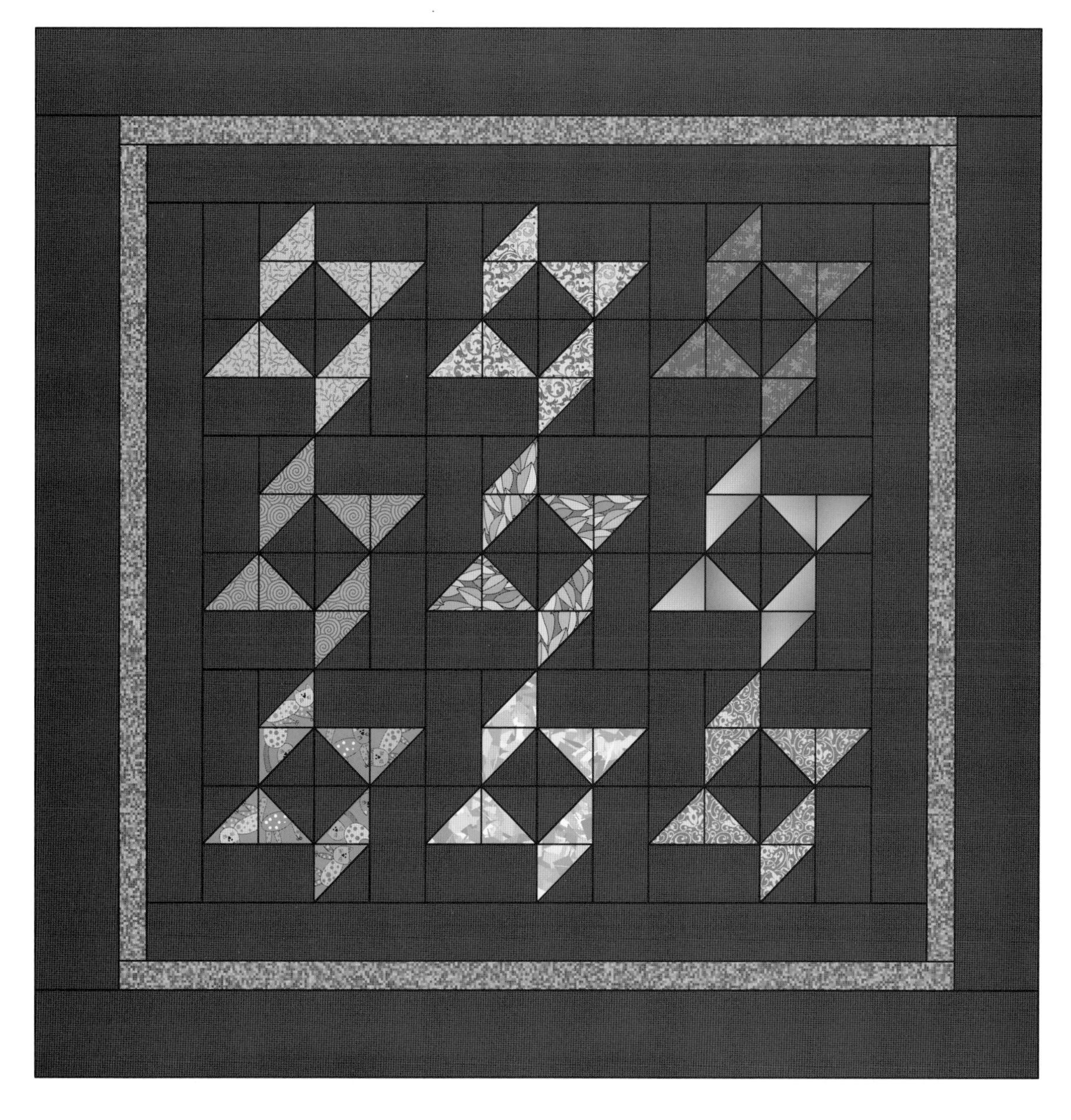

Assorted bright prints

Black solid

Multicolor print

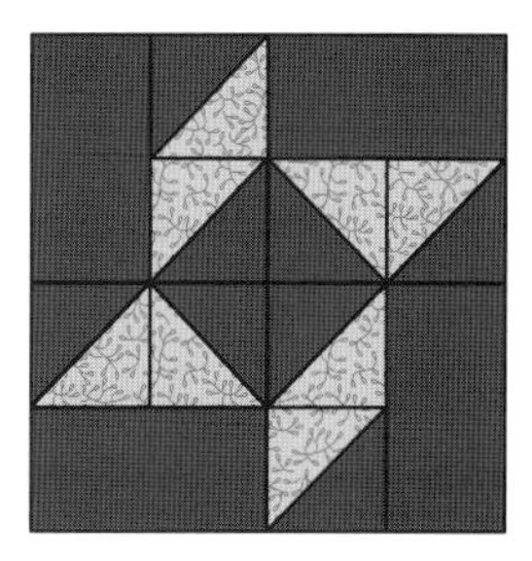

Rhumba Star
Make 9.

Rhumba Star is a great little quilt for starters. Using simple shapes, the dancing stars go together easily and quickly. If you would like to use a different background print for each block instead of the black solid, use one 8" x 18" background piece for each block. Since each block is made from a different set of prints, you can easily increase your quilt to any size.

Rhumba Star will provide you with experience in:

Cutting Squares (page 12)
Cutting Rectangles (page 12)
Bias Strip Piecing (page 28)

Materials: 42"-wide fabric

1 piece, 8" x 14", each of 9 bright prints
1½ yds. black solid
¼ yd. bright multicolor print
1¼ yds. for backing
⅜ yd. for binding

Cutting

From each bright print, cut:

2 squares, each 6" x 6", for bias strip piecing

From the black solid, cut:

3 strips, each 6" x 42"; cut strips into 18 squares, each 6" x 6", for bias strip piecing

4 strips, each 2½" x 42"; cut strips into 36 rectangles, each 2½" x 4½"

4 strips, each 2½" x 42", for the inner border

4 strips, each 3½" x 42", for the outer border

From the bright multicolor print, cut:

4 strips, each 1½" x 42", for the middle border

From the fabric for binding, cut:

4 strips, each 2" x 42"

Assembling the Blocks

The following instructions are for making one block.

1. Pair each bright square with a black square, right sides together. Cut the squares once diagonally; then cut a set of 2½"-wide bias strips from each pair of triangles using a regular rotary ruler. (Or, cut strips at the 2¾" mark using the Bias Stripper.) Sew the pairs of bias strips into strip units along their long bias edges. Press. Cut 2 bias squares, each 2½" x 2½", from each strip unit. Half will have their seam allowances pressed to the bright print and half to the black solid.

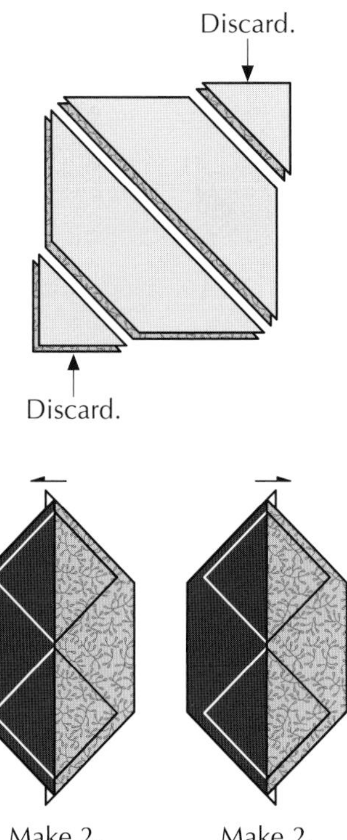

Make 2. Make 2.

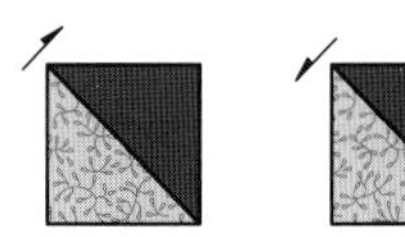

Cut 4 each.

2. Sew the bias squares as shown to make 2 of Unit A and 2 of Unit B. Make sure the seam allowances are oriented in the proper direction as shown.

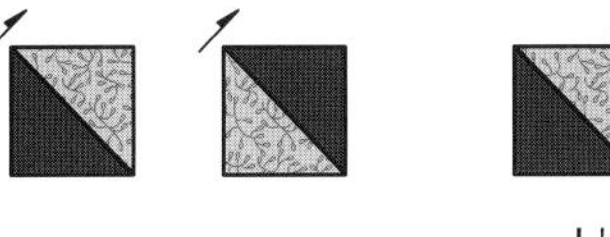

Unit A
Make 2.

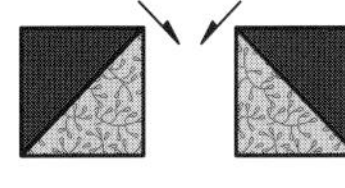

Unit B
Make 2.

3. Sew a black rectangle to each of the units made in step 2.

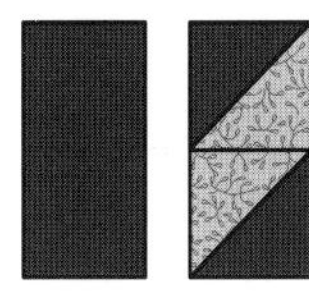

Make 2.

Make 2.

4. Arrange and sew the units into rows as shown. Join the rows to complete a block.

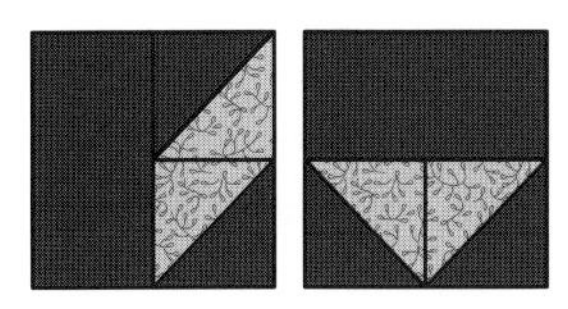
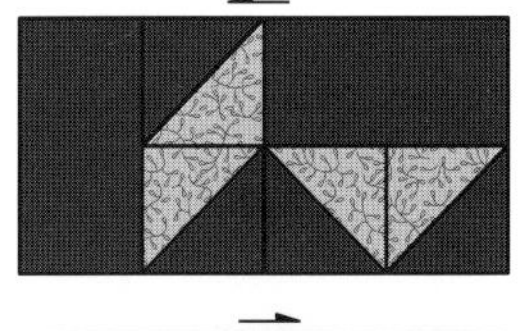
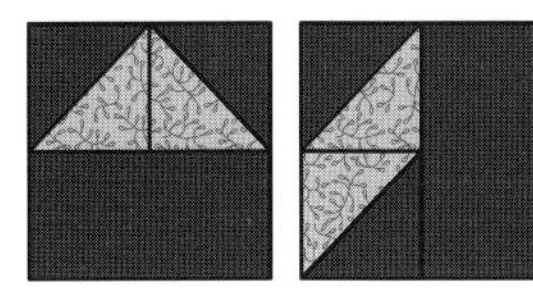
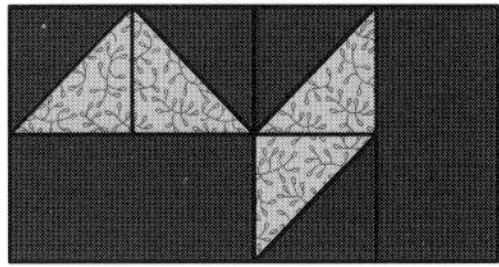

Repeat steps 1–4 with the remaining bright prints to make 8 more Rhumba Star blocks.

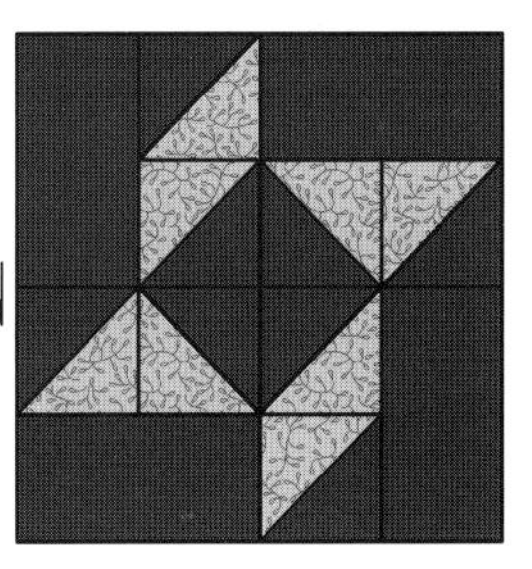

Assembling and Finishing the Quilt

1. Arrange and sew the blocks into 3 rows of 3 blocks each, placing the final block seams as shown. Be sure to have all your stars dancing in the same direction. Join the rows. Press.

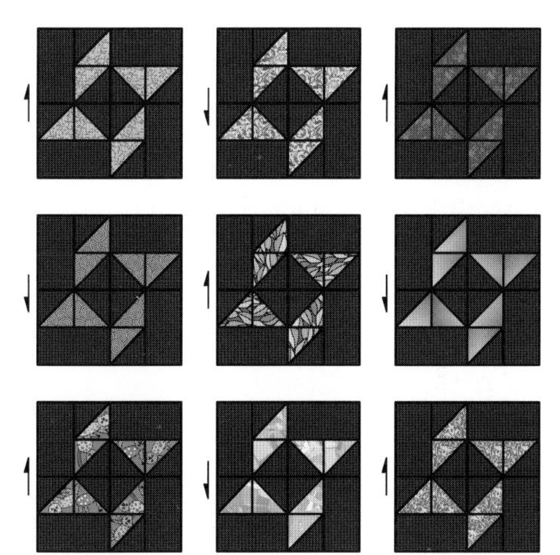

2. Referring to "Borders" on pages 60–61, measure the quilt top and cut the black inner border strips to size. Sew the strips to the quilt top. Repeat for the multicolor-print middle border and the black outer border.
3. Layer the completed quilt top with batting and backing; baste. Quilt as desired. Bind the edges of the quilt. Add a label.

Legacy

Color photo on page 36

Finished Quilt Size: 34" x 34"
Finished Block Size: 8" x 8"

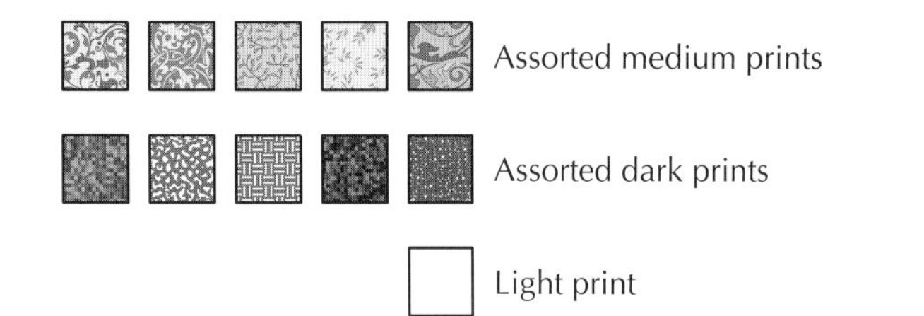

Legacy
Make 9.

Legacy is an easily constructed quilt that uses straight-grain strip piecing. The squares in the border are set on-point and should measure evenly across the diagonal to fit standard sizes. To do this, you can use one of two tools, a paper cutting guide or the Bias Stripper ruler. Directions are provided for both methods.

The blocks are made two at a time from five different fabric combinations. Since there are only nine blocks in the quilt, you can use the left-over block on the back to label the quilt, or you can make it into a pillow or other project.

Legacy will provide you with experience in:

Cutting Half-Square Triangles (page 12)
Cutting Quarter-Square Triangles (page 13)
Cutting On-Point Squares (page 16)
Straight-Grain Strip Piecing (page 26)

Materials: 42"-wide fabric

1 piece, 8" x 21", each of 5 medium prints
1 piece, 10" x 21", each of 5 dark prints
1½ yds. light print
1¼ yds. for backing
⅜ yd. for binding

Cutting

All cutting dimensions are for a regular ruler unless you are instructed to use either the Bias Stripper or the paper cutting guides on page 45.

The cutting directions for the medium and dark prints are for making two blocks at a time. Decide which medium print will go with each dark print and keep all the pieces cut from those two prints together. Repeat the cutting directions for the medium and dark fabrics five times for a total of ten blocks.

From each medium print, cut:

3 strips, each 20" long. Cut the strip width at the 2" mark, using the Bias Stripper or the square cutting guide. Cut 2 of the 20"-long strips into 4 pieces, each 10" long, for strip piecing. From the remaining strip, cut 7 squares at the 2" mark, using the Bias Stripper or the square cutting guide. Set these squares aside for the pieced border.
1 square, 1½" x 1½", for sashing squares

From each dark print, cut:

4 strips, each 20" long. Cut the strip width at the 2" mark, using the Bias Stripper or the square cutting guide. Cut 3 of the 20"-long strips into 5 pieces, each 10" long, for strip piecing. From the remaining strip pieces, cut 11 squares at the 2" mark, using the Bias Stripper or the square cutting guide. Set aside 4 squares for block piecing and 7 for the pieced border.

The cutting directions for the light print are for making all 10 blocks.

From the light print, cut:

3 strips, each 42" long. Cut the strip width at the 2" mark, using the Bias Stripper or the square cutting guide. Cut 2 of the 42"-long strips into 5 pieces, each 10" long, for strip piecing. From the remaining strip lengths, cut 10 rectangles, each at the 6" mark, using the Bias Stripper or the rectangle cutting guide.

5 strips, each 3¼" x 42"; cut strips into 60 squares, each 3¼" x 3¼". Cut each square twice diagonally to make 240 quarter-square triangles.

2 strips, each 1⅞" x 42"; cut strips into 28 squares, each 1⅞" x 1⅞". Cut each square once diagonally to make 56 half-square triangles.

3 strips each 1½" x 42"; cut strips into 12 sashing strips, each 1½" x 8½"

4 strips, each 2½" x 42", for the inner border

From the fabric for binding, cut:

4 strips, each 2" x 42"

Assembling the Blocks

The following instructions are for making a pair of blocks, one from each dark/medium combination.

1. Assemble Strip Units I, II, and III as shown. Cut each strip unit into 4 segments at the 2" mark, using the Bias Stripper or square paper cutting guide.

Strip Unit I

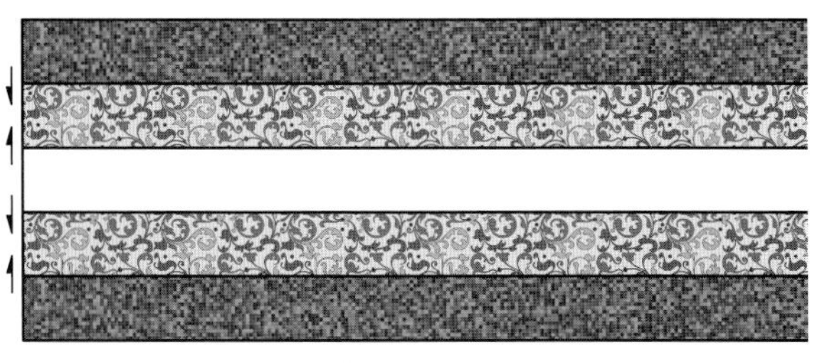

Strip Unit II

Strip Unit III

2. Sew segments from Strip Unit III to each end of a light rectangle. Make 2.

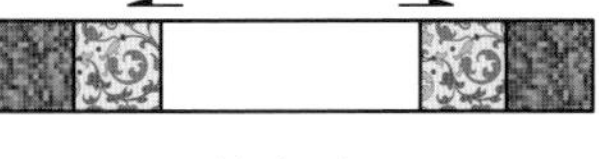

Make 2.

3. Arrange 2 segments from Strip Unit I, 2 segments from Strip Unit II, 1 unit from step 2, 2 dark squares, and 12 light quarter-square triangles as shown. Sew the units into rows. Join the rows. Press the seams away from the center. Sew a light 1⅞" half-square triangle to each corner to complete the block. Make 2 blocks.

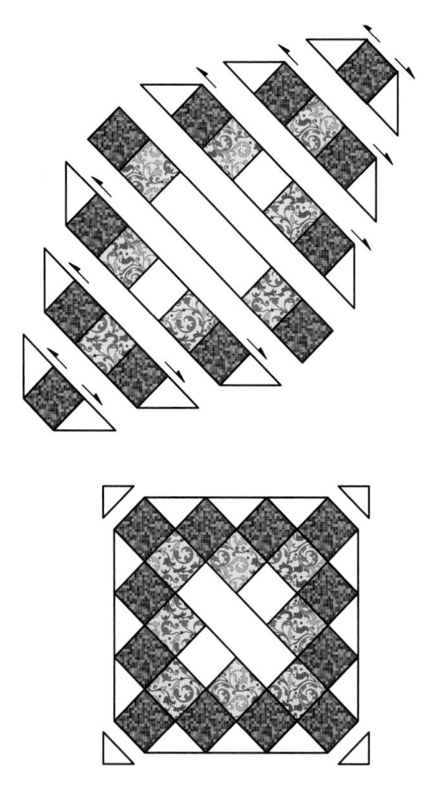

Repeat steps 1–3 to make 4 more pairs of blocks with the remaining medium/dark combinations. Set aside 1 block for the back or another project.

Assembling and Finishing the Quilt

1. Arrange and sew the 9 blocks, 12 sashing strips, and 4 sashing squares into rows as shown in the quilt plan. Press the seams toward the sashing strips. Join the rows. Press. You will have 1 sashing square left over.
2. Referring to "Borders" on pages 60–61, measure the quilt top and cut the light inner border strips to size. Sew the strips to the quilt top. Press seams toward the border strips.
3. Sew a light quarter-square triangle to opposite sides of a dark or medium square as shown. Make 56 side units.

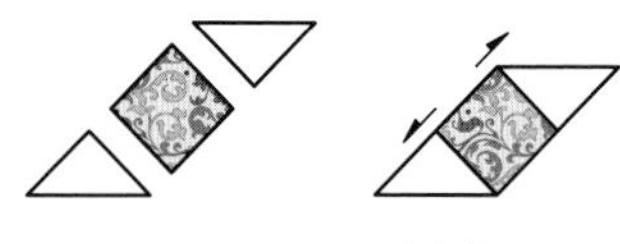

Make 56.

4. Sew a light quarter-square triangle and 2 light half-square triangles to a dark or medium square as shown. Make 8 corner units. You will have 6 dark or medium squares left over.

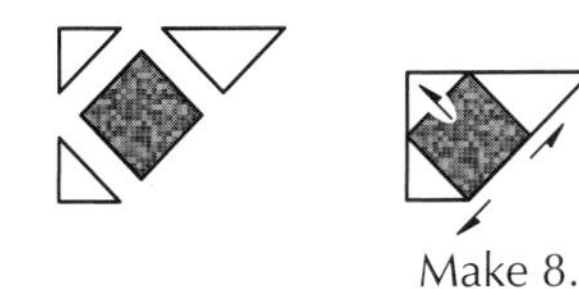

Make 8.

5. Sew 13 side units together as shown. Add a corner unit to each end of the strip. Make 2 side borders. Sew the side borders to opposite sides of the quilt.

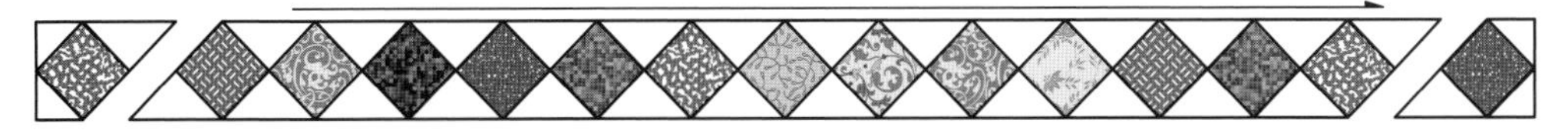

6. Sew 15 side units together as shown. Add a corner unit to each end of the strip. Make 2 long borders. Sew them to the top and bottom of the quilt.

7. Layer the completed quilt top with batting and backing; baste. Quilt as desired. Bind the edges of the quilt. Add a label.

Square cutting guide

Rectangle cutting guide

Seven Sisters

Color photo on page 33

Finished Quilt Size: 30" x 26¼"

This variation of Seven Sisters is unique because I replaced the usual 60° diamonds found in the traditional pattern with two equilateral triangles. By so doing, I eliminated all the set-in seams required in the original block. The quilt is a tessellation (repeat) of one particular shape—an equilateral triangle. There are no blocks to make. The triangles are cut, then arranged and sewn into long rows to make the overall design.

I chose two color families, dark blue and gold, then selected many prints from each color family to create a scrappy look. If you prefer to make each star from a different fabric, you'll need twelve triangles for each of the seven stars and sixty-six triangles for the background.

Seven Sisters will provide you with experience in:

Cutting Equilateral Triangles using the mark, stack, and cut method (page 15)

Materials: 42"-wide fabric

1 yd. total of assorted dark blue prints
1⅜ yds. total of assorted gold prints
1 yd. for backing
⅜ yd. for binding

Cutting

Make a plastic or cardboard template of the equilateral triangle below.

From the assorted dark blue prints, cut:
66 squares, each 4¼" x 4¼". Stack them in 11 groups of 6. Using the template, draw a triangle on the top fabric in each stack. Cut on the 3 lines through all layers in each stack. Be careful not to shift the fabrics in the stack as you cut. Cut a total of 66 triangles.

From the assorted gold prints, cut:
84 squares, each 4¼" x 4¼". Follow the same procedure as above to mark, stack, and cut 84 triangles.

From the fabric for binding, cut:
3 strips, each 2¼" x 42"*

*These strips are wider than I normally use. The wider binding fits better around the wide-angled corners of this quilt.

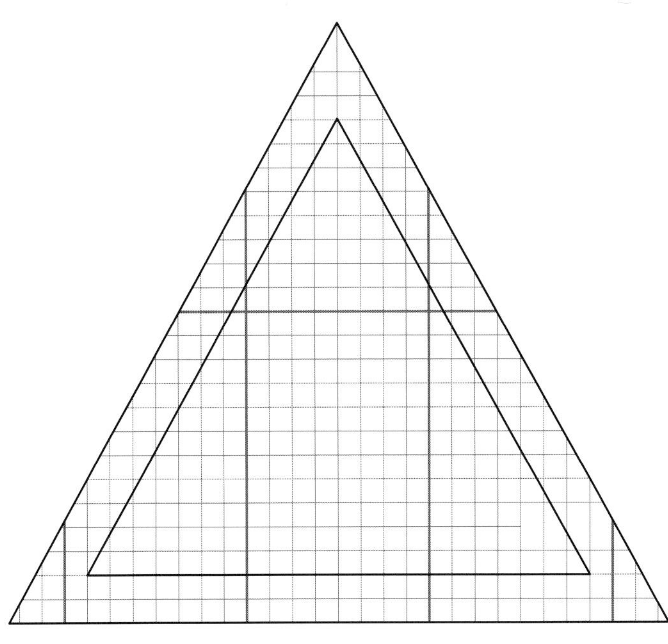

Triangle Template

Assembling and Finishing the Quilt

1. Arrange and sew the triangles into horizontal rows as shown. Press the seams in opposite directions from row to row so they will butt later. There are a lot of bias edges, so press gently to avoid distorting the triangles.

2. Join the rows, butting the diagonal seams at each intersection. Press gently.
3. Layer the completed quilt top with batting and backing; baste. Quilt as desired. Bind the edges of the quilt. The process for binding an octagonal quilt is the same as an ordinary quilt. Turn the corner as shown below. The turning pleat will be a little smaller than usual. Add a label.

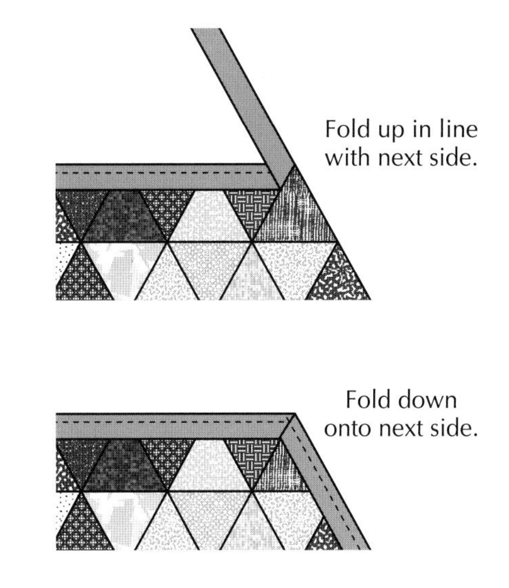

Never Too Late

Color photo on page 34

Finished Quilt Size: 87⅞" x 87⅞"
Finished Block Size: 12" x 12"

Purple print

Green print

Red print

Beige print

Floral print

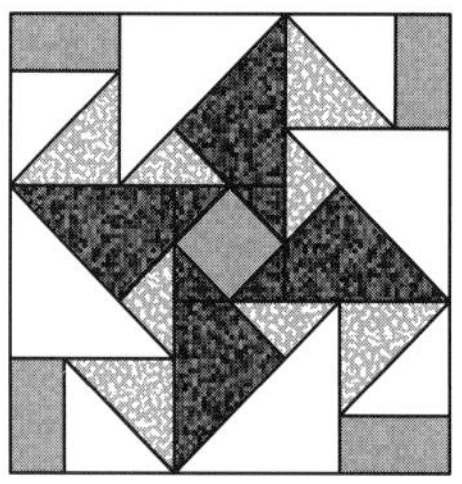

Whirligig
Make 25.

This delightful Whirligig block is set on-point to enhance the spin of the rectangles where they meet at the corners of the blocks. There is a lot of movement in this quilt. Use your boldest fabric for the large spinning triangles.

Never Too Late will provide you with experience in:

Cutting Rectangles (page 12)
Cutting Half-Square Triangles (page 12)
Cutting Side-Setting and Corner-Setting Triangles (page 13)
Making Pieced Squares (page 23)
Making Square-within-a-Square Units (page 23)
Modifying Triangles (page 25)

Materials: 42"-wide fabric

2⅛ yds. purple print
2⅛ yds. green print
1 yd. red print
2 yds. beige print
2¾ yds. floral print
8⅛ yds. for backing
⅝ yd. for binding

Cutting

From the purple print, cut:

8 strips, each 5⅜" x 42"; cut strips into 50 squares, each 5⅜" x 5⅜". Cut squares once diagonally to yield 100 half-square triangles.

5 strips, each 2" x 42"; cut strips into 100 squares, each 2" x 2"

7 strips, each 2" x 42", for the inner border

From the green print, cut:

10 strips, each 3⅞" x 42"; cut strips into 100 squares, each 3⅞" x 3⅞"

8 strips, each 3½" x 42", for the middle border

From the red print, cut:

3 strips, each 3½" x 42"; cut strips into 25 squares, each 3½" x 3½"

5 strips, each 3½" x 42"; cut strips into 100 rectangles, each 2" x 3½"

From the beige print, cut:

8 strips, each 5⅜" x 42"; cut strips into 50 squares, each 5⅜" x 5⅜". Cut squares once diagonally to yield 100 half-square triangles.

5 strips, each 3⅞" x 42"; cut strips into 50 squares, each 3⅞" x 3⅞"

From the floral print, cut:

3 squares, each 19" x 19"; cut squares twice diagonally to yield 12 side triangles

2 squares, each 10" x 10"; cut squares once diagonally to yield 4 corner triangles

8 strips, each 6" x 42", for the outer border

From the fabric for binding, cut:

9 strips, each 2" x 42"

Assembling the Blocks

The following instructions are for making all 25 blocks. To save time, chain sew the pieces together (see page 60).

1. Draw a diagonal line on the wrong side of each 2" purple square. Lay a marked square on the corner of a red square, right sides together, as shown.

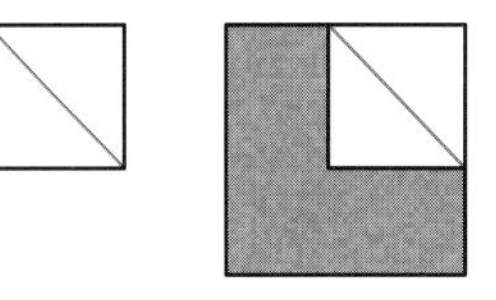

2. Stitch on the marked line. Fold the corner of the small square over itself to check the accuracy of the stitching. Readjust if necessary. Trim as desired (see page 22) and press the purple triangle over the corner. Repeat for each of the 4 corners of the square. Make 25 pieced centers.

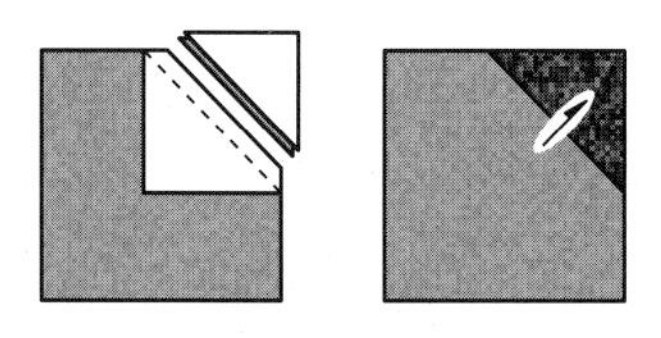

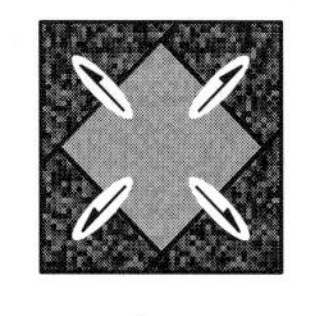

Make 25.

3. Draw a diagonal line on the wrong side of 50 green squares. Cut each marked square on the opposite diagonal to make 100 marked half-square triangles.

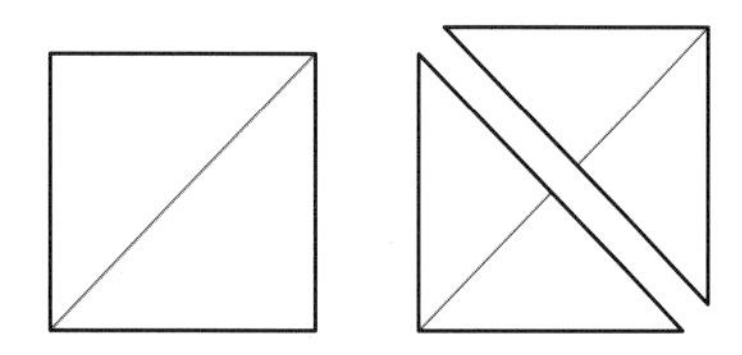

4. Lay a marked green triangle on the corner of a purple triangle, right sides together, as shown. Carefully align the points and raw edges. Stitch on the marked line. Fold the green triangle over itself to check the accuracy of the stitching. Readjust the seam, if necessary, so the top layer matches the bottom. Trim as desired (see page 22). Make 100 pieced triangles.

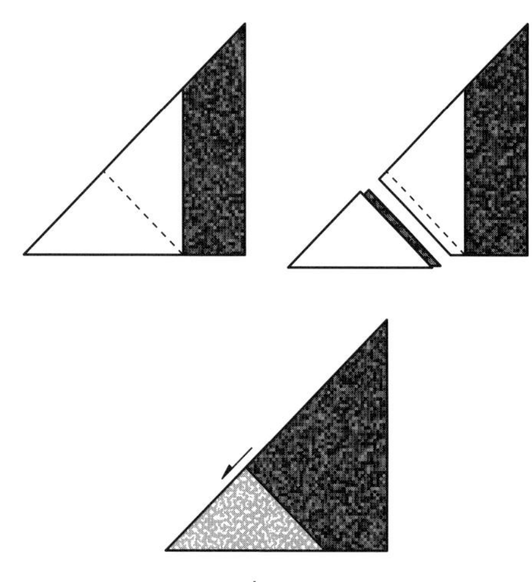

Make 100.

5. Sew a pieced triangle to a beige triangle. Make 100 large pieced squares.

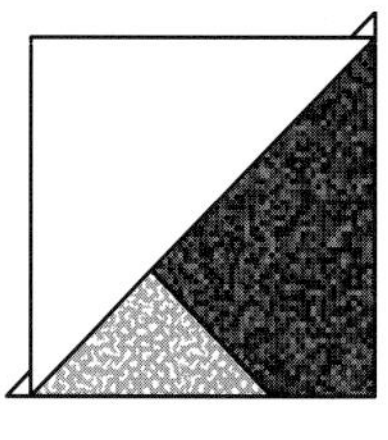

Make 100.

6. Draw a diagonal line on the wrong side of each beige square. Pair a beige square with a green square, right sides together. Stitch ¼" from the marked line on both sides. Cut on the marked line and press. Make 100 small pieced squares.

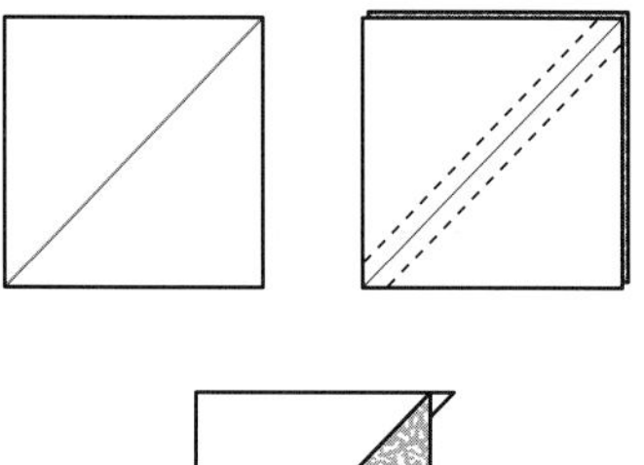

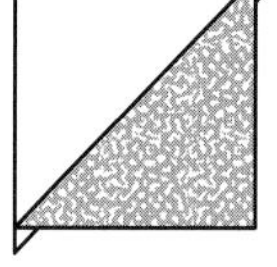

Make 100.

7. Sew a small pieced square to a red rectangle as shown. Make 100 rectangle units.

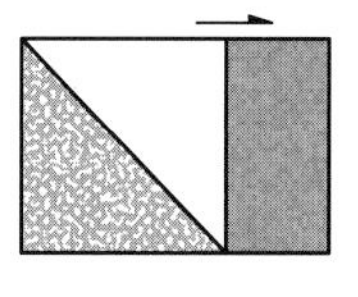

Make 100.

8. Sew a rectangle unit to a large pieced square as shown. Make 100 block units.

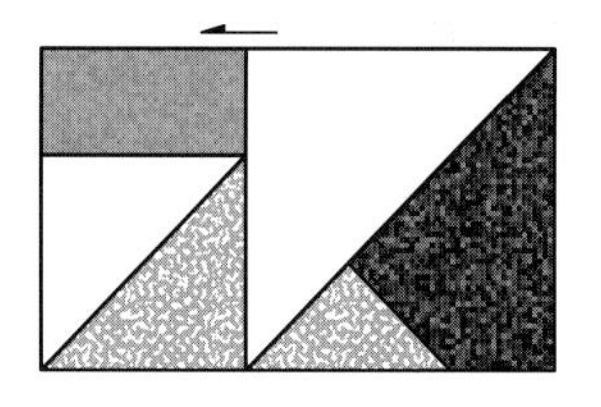

Make 100.

9. Sew a block unit to a pieced center, using a partial seam as shown. Sew slightly past the center of the pieced square and backstitch. Stop and remove the unit from the machine.

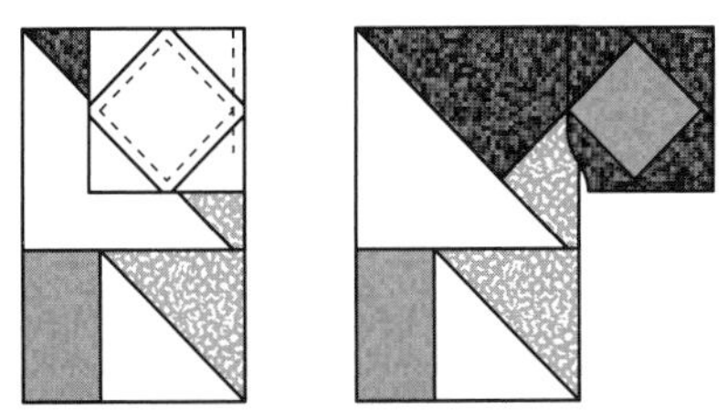

10. Sew the remaining block units to the center unit as shown, stitching each seam from edge to edge. Press the seams away from the center unit. When all are attached, finish stitching the last portion of the first seam to complete the block.

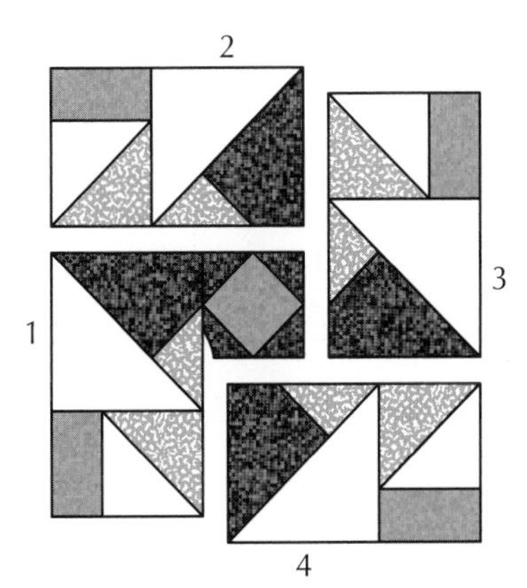

Make 25.

Assembling and Finishing the Quilt

1. Arrange and sew the blocks and side triangles into diagonal rows as shown. Press the seams in opposite directions from row to row. Join the rows. Add the 4 corner triangles last. Press.

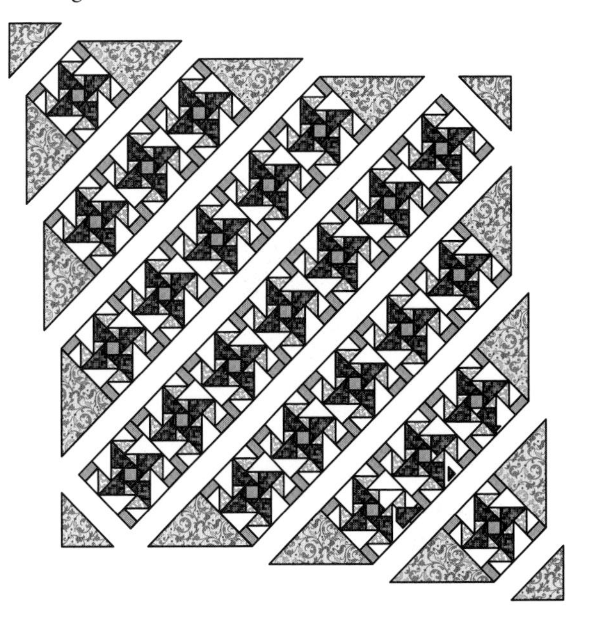

2. Referring to "Borders" on pages 60–61, measure the quilt top and cut the inner border strips to size. Sew the strips to the quilt top. Repeat for the middle and outer borders.
3. Layer the completed quilt top with batting and backing; baste. Quilt as desired. Bind the edges of the quilt. Add a label.

CHRISTMAS SPLENDOR

Color photo on page 31

Finished Quilt Size: 66" x 66"
Finished Block Size: 12" x 12"

Green print

Red print

Beige print

Light print

Optical Illusion
Make 16.

Christmas Splendor is made from Optical Illusion blocks, which when set together straight without sashing, create a wonderful circular motion. Of course, you do not have to use Christmas prints to make this quilt. Take a look at "Walrus in the Mirror" and "Delta Zeta Girls" on page 32 to see how two friends made their quilts, using different color schemes.

Christmas Splendor will provide you with experience in:

Cutting Half-Square Triangles (page 12)
Cutting Quarter-Square Triangles (page 13)
Cutting Half-Rectangle Triangles (page 18)
Nubbing (page 21)
Making Pieced Squares (page 23)

Materials: 42"-wide fabric

2½ yds. green print
⅞ yd. red print
⅜ yd. beige print
1½ yds. light print
½ yd. plaid
4¼ yds. for backing
⅝ yd. for binding

Cutting

Use the paper cutting guide on page 55 for cutting the rectangles.

From the green print, cut:

2 strips, each 5¼" x 42"; cut strips into 16 squares, each 5¼" x 5¼". Cut squares twice diagonally to yield 64 quarter-square triangles.

5 strips, each 42" long, cut to the short width of the rectangle cutting guide. Cut strips into 32 rectangles, using the length of the rectangle cutting guide.

3 strips, each 2⅞" x 42"; cut strips into 32 squares, each 2⅞" x 2⅞"

6 strips, each 8" x 42", for the outer border

From the red print, cut:

2 strips, each 5¼" x 42"; cut strips into 16 squares, each 5¼" x 5¼". Cut squares twice diagonally to yield 64 quarter-square triangles.

5 strips, each 42" long, cut to the short width of the rectangle cutting guide. Cut strips into 32 rectangles, using the length of the rectangle cutting guide.

From the beige print, cut:

3 strips, each 2⅞" x 42"; cut strips into 32 squares, each 2⅞" x 2⅞"

From the light print, cut:

4 strips, each 4⅞" x 42"; cut strips into 32 squares, each 4⅞" x 4⅞". Cut squares once diagonally to yield 64 half-square triangles.

10 strips, each 42" long, cut to the short width of the rectangle cutting guide. Cut strips into 64 rectangles, using the length of the rectangle cutting guide.

From the plaid fabric, cut:

5 strips, each 2" x 42", for the inner border

From the fabric for binding, cut:

7 strips, each 2" x 42"

Assembling the Blocks

The following instructions are for making all 16 blocks. To save time, chain sew the pieces together (see page 60).

1. With right sides up, cut the green and red rectangles in half on the diagonal as shown. You will be cutting the 2 prints on opposite diagonals. With right sides up, cut 32 light rectangles on 1 diagonal and 32 on the opposite diagonal as shown.

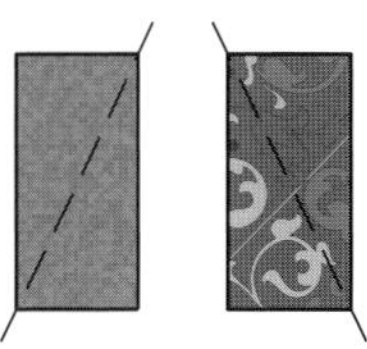

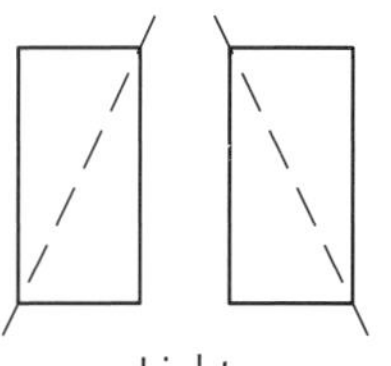

2. Using the nubbing guide on page 55, nub each half-rectangle triangle.

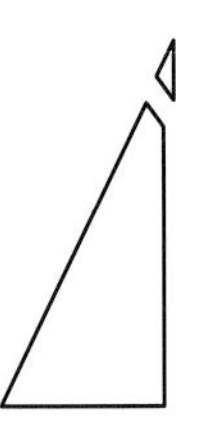

3. Sew the light half-rectangle triangles to the green and red half-rectangle triangles as shown. Make 64 pieced rectangles of each color.

Make 64.

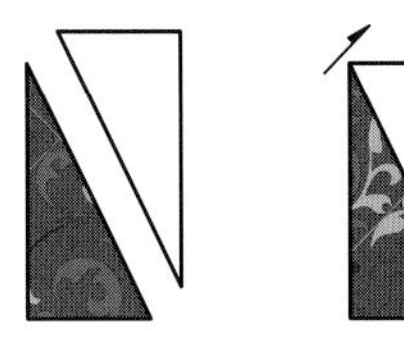

Make 64.

4. Sew a green pieced rectangle to a red pieced rectangle. Make 64.

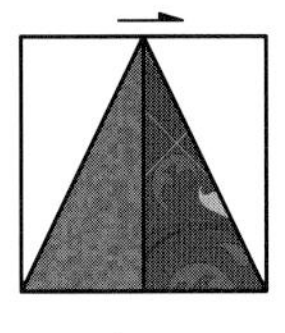

Make 64.

5. Sew a green quarter-square triangle to a red quarter-square triangle. Make 64.

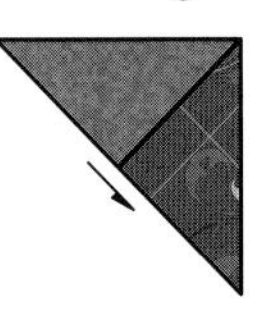

Make 64.

6. Sew a light half-square triangle to a quarter-square triangle pair from step 5. Make 64.

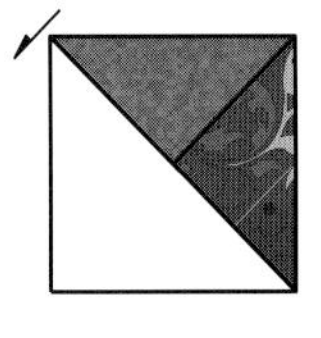

Make 64.

7. Draw a diagonal line on the wrong side of each beige square. Place it on a green square, right sides together. Sew ¼" from the marked line on both sides. Cut on the marked line and press to make 2 pieced squares. Make a total of 64 pieced squares.

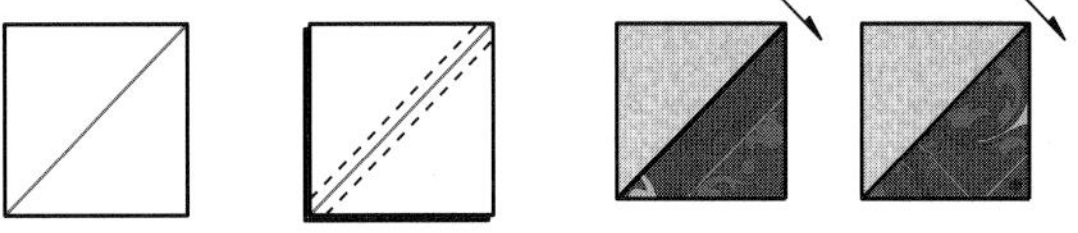

Make 64.

8. Sew 4 pieced squares from step 7 into a pinwheel as shown. Pull out a few threads from the last seam to press the seams in a clockwise direction. Make 16 pinwheels.

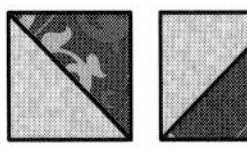

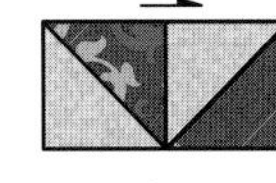

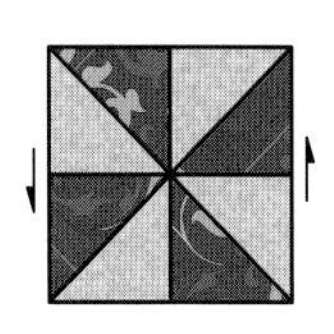

Make 16.

9. Arrange and sew the units into rows as shown. Join the rows to complete the block.

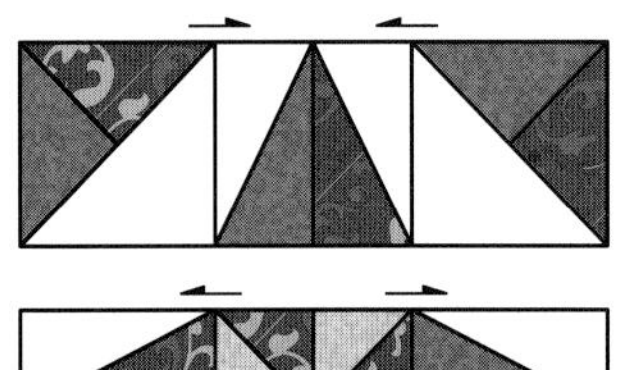

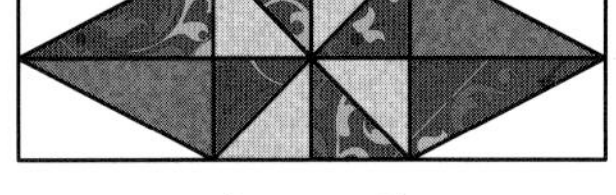

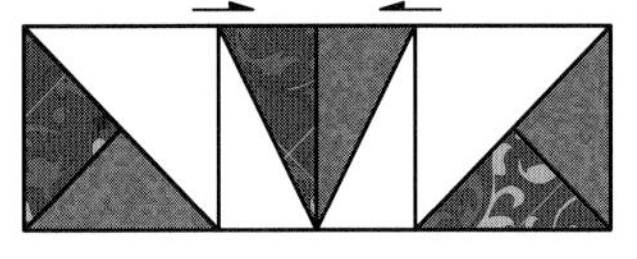

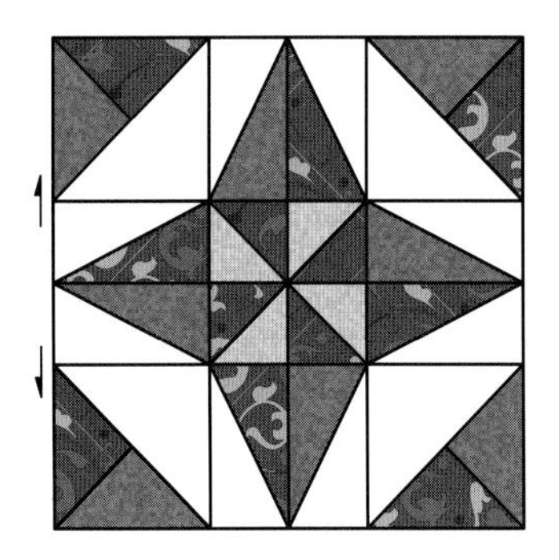

Make 16.

Assembling and Finishing the Quilt

1. Arrange and sew the blocks into 4 rows of 4 blocks each, placing the final block seams as shown. Join the rows. Press.

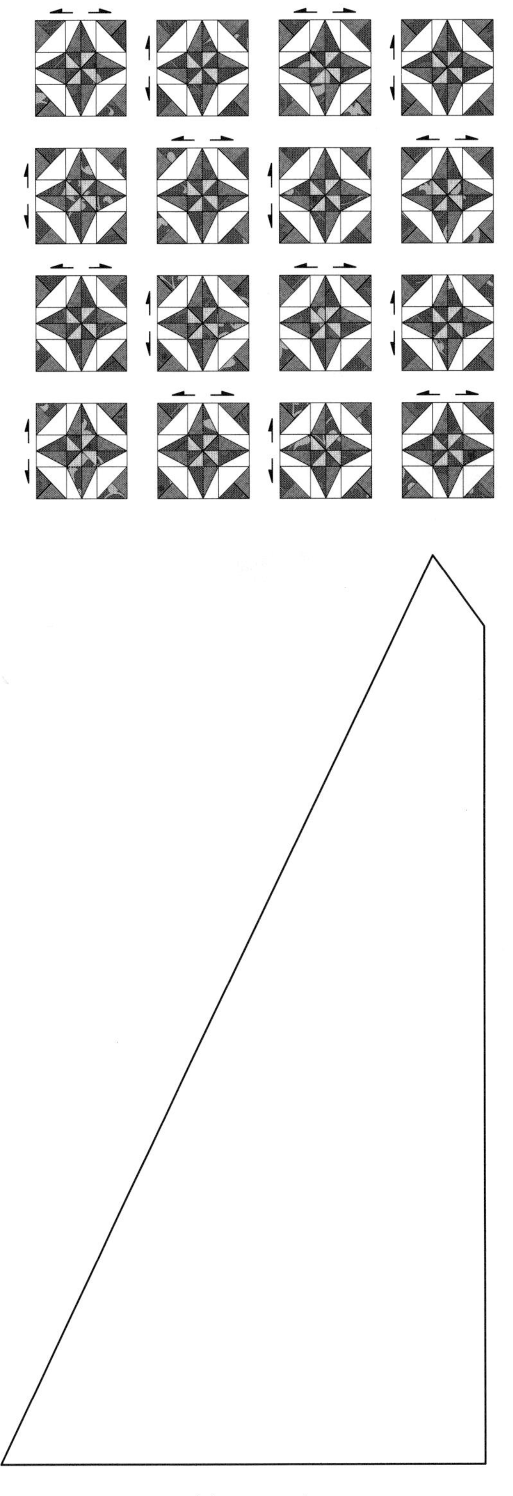

Nubbing Guide

2. Referring to "Borders" on pages 60–61, measure the quilt top and cut the inner border strips to size. Sew the strips to the quilt top. Repeat for the outer border.
3. Layer the completed quilt top with batting and backing; baste. Quilt as desired. Bind the edges of the quilt. Add a label.

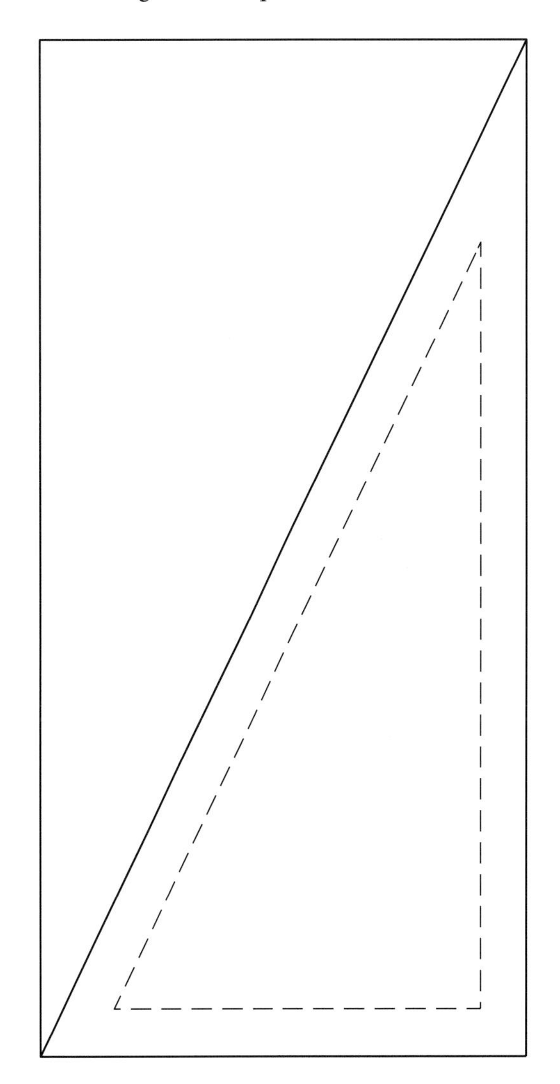

Rectangle Cutting Guide

Daylilies Party at Night

Color photo on page 29

Finished Quilt Size: 64" x 64"
Finished Block Size: 12" x 12"

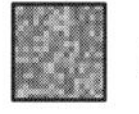
Dark bright

Medium bright

Dark green

Medium green

Navy blue

Multicolor print

Lilies
Make 9.

There are many different prints used in the flowers of this lovely Lily pattern. Although each block is made from a different combination of prints, the colors were all pulled from the multicolor print used in the outer border. Each block is made from two shades of the same bright color and two green prints.

Deb mitered the striped inner border of her quilt. You may or may not choose to miter when using a striped print. The pattern directions are for a border with straight-cut corners.

Daylilies Party at Night will provide you with experience in:

Cutting Squares (page 12)
Cutting Half-Square Triangles (page 12)
Cutting Quarter-Square Triangles (page 13)
Cutting Side-Setting and Corner-Setting Triangles (page 13)
Cutting Diamonds (page 18)
Making Pieced Triangles (page 24)

Materials: 42"-wide fabric

1 piece, 6" x 18", each of 9 dark bright prints
1 piece, 6" x 18", each of 9 medium bright prints

1 piece, 6" x 12", each of 9 dark green prints
1 square, 5" x 5", each of 9 medium green prints

2⅝ yds. navy blue print
⅞ yd. multicolor print
½ yd. striped print
4¼ yds. for backing
½ yd. for binding

Cutting

Use the paper cutting guide on page 59 or the 2⅛" mark on the Bias Stripper to cut the diamonds.

From each dark and medium bright print, cut:
8 diamonds

From each dark green print cut:
2 squares, each 3⅞" x 3⅞"; cut squares once diagonally to yield 4 half-square triangles

From each medium green print, cut:
4 squares, each 2" x 2"

From the navy blue print, cut:
3 strips, each 2⅝" x 42"; cut strips into 36 squares, each 2⅝" x 2⅝"

2 strips, each 4¼" x 42"; cut strips into 18 squares, each 4¼" x 4¼". Cut squares twice diagonally to yield 72 quarter-square triangles.

6 strips, each 2¼" x 42"; cut strips into 36 strips, each 2¼" x 5⅝"

4 squares, each 12½" x 12½", for alternate plain blocks

2 squares, each 19" x 19"; cut squares twice diagonally to yield 8 side-setting triangles

2 squares, each 10" x 10"; cut squares once diagonally to yield 4 corner-setting triangles

6 strips, each 2" x 42", for the middle border

From the multicolor print, cut:
6 strips, each 4" x 42", for the outer border

9 squares, each 2¼" x 2¼", for block centers

From the striped print, cut:
6 strips, each 2" x 42", for the inner border

From the fabric for binding, cut:
7 strips, each 2" x 42"

Assembling the Blocks

The following instructions are for making one block.

1. On the wrong side of each diamond, 8 navy blue quarter-square triangles, and 4 navy blue 2⅝" squares, draw crosshairs on the ¼" seam intersection at each corner.

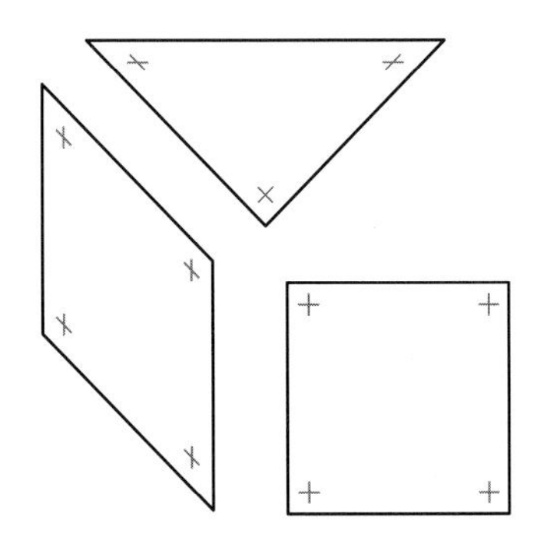

2. Sew 2 medium bright diamonds together from crosshair to crosshair, backstitching at the start and finish of the seam. Set in the 2 seams of a navy blue square by pinning and sewing the first seam from the center out. Sew from crosshair to crosshair, backstitching at the start and finish of the seam. Pin and sew the second seam in the same fashion. Make 8 diamond pairs.

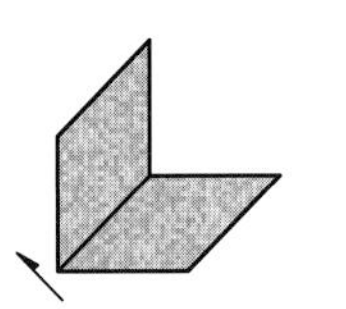
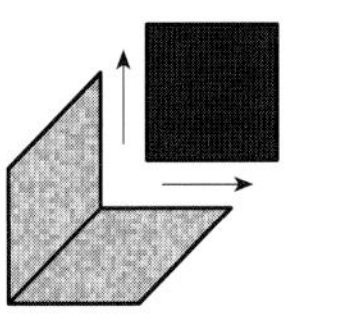
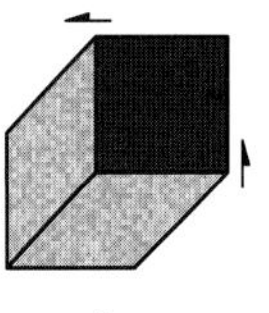

Make 8.

3. Sew a darker diamond to either side of the diamond pair from step 2, pinning and sewing from corner to corner as before. Set in 2 navy blue quarter-square triangles in the same fashion as the square in step 2. Make 4 diamond units.

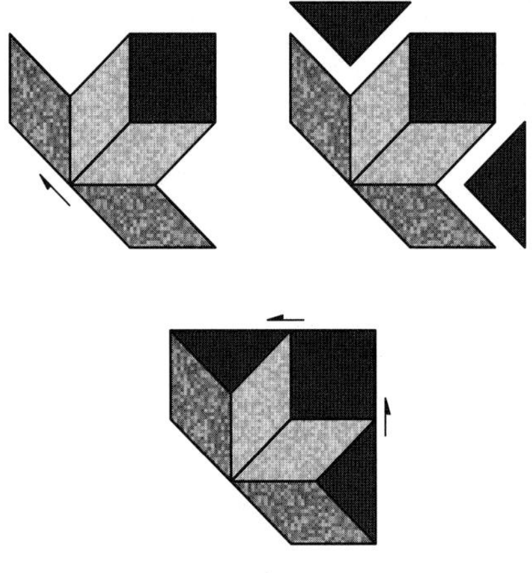

Make 4.

4. Draw a pencil line on the wrong side of a medium green square. Place it right sides together on a dark green half-square triangle as shown. Pin in place and stitch on the line. Trim the excess ¼" from the stitching line. Make 4 pieced triangles.

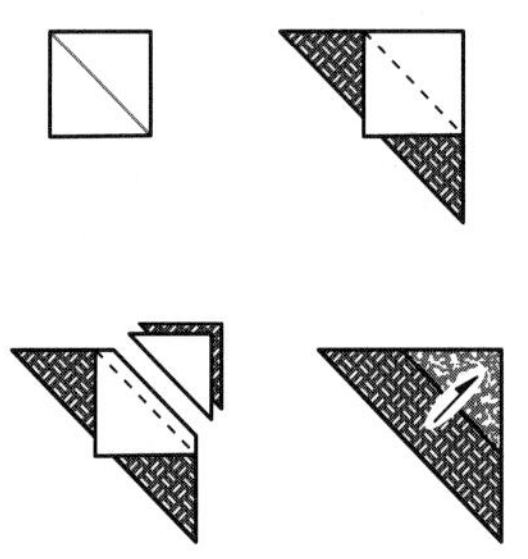

Make 4.

5. Sew a pieced triangle to a diamond unit. Make 4 lily units.

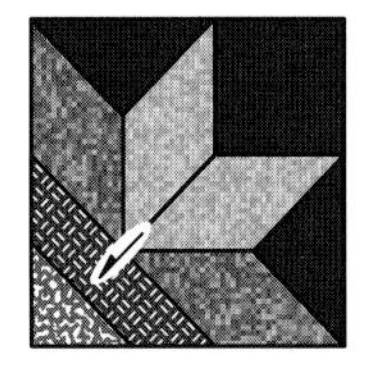

Make 4.

6. Arrange the lily units, the 2¼" x 5⅝" navy blue strips, and the 2¼" multicolor print square as shown. Sew the units into rows. Join the rows to complete a Lily block.

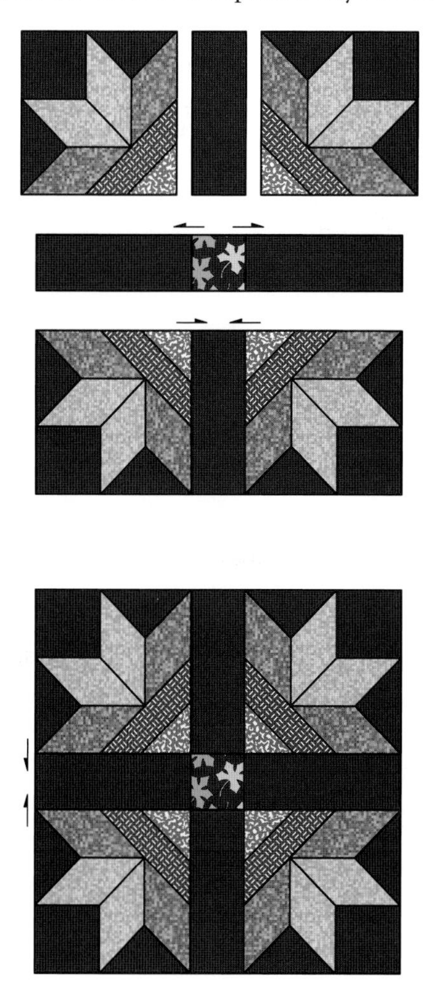

Make 9.

Repeat steps 1–6 to make 8 more Lily blocks from the remaining bright and green prints.

Assembling and Finishing the Quilt

1. Arrange and sew the Lily blocks, the large navy blue squares, and the side triangles into diagonal rows as shown. Press the seams toward the navy blue squares and side triangles. Join the rows. Add the 4 corner triangles last. Press.

2. Referring to "Borders" on pages 60–61, measure the quilt top and cut the inner border strips to size. Sew the strips to the quilt top. Repeat for the middle and outer borders.
3. Layer the completed quilt top with batting and backing; baste. Quilt as desired. Bind the edges of the quilt. Add a label.

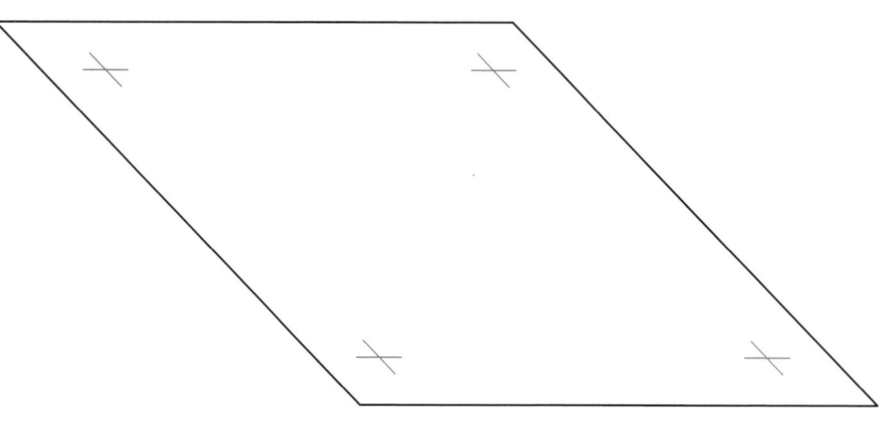

Diamond Cutting Guide

Assembling and Finishing Techniques

Basic Stitching

Most, but not all, machine piecing is done from raw edge to raw edge. Backstitching is normally unnecessary because each seam will be crossed by another, securing the stitches in the process. Sometimes, however, seam allowances are left free, such as in the Lily block (page 58), and you must backstitch these seams.

In general, though, to join rotary-cut pieces, place them right sides together. Carefully and accurately align the raw edges. Sew slowly and accurately, keeping the edges aligned. For more extensive information on machine piecing, see my book *A Perfect Match.*

Matching Intersections

The easiest and most accurate way to match intersections is to press seam allowances in opposite directions. Each of the seam allowances forms a ridge, and these ridges can be pushed tightly against each other. This is called "butting the seams." Butting also applies to diagonal seams.

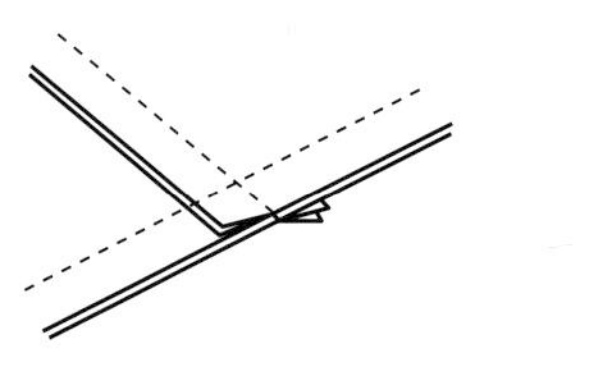

Butt straight seams.

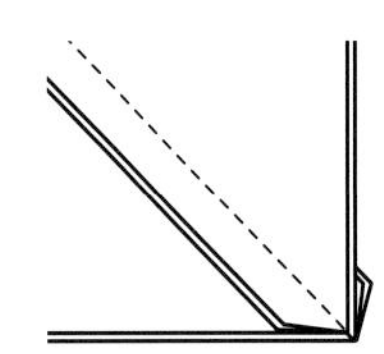

Butt diagonal seams.

Chain Sewing

Chain sewing is an assembly-line approach to stitching. The idea is to save time and increase accuracy by sewing as many seams as possible, one right after the other, rather than stopping and starting after each unit is sewn. When you've finished sewing a set of seams, you should have a long "kite tail" of stitched units connected by small twists of thread. Clip the units apart and press according to the directions.

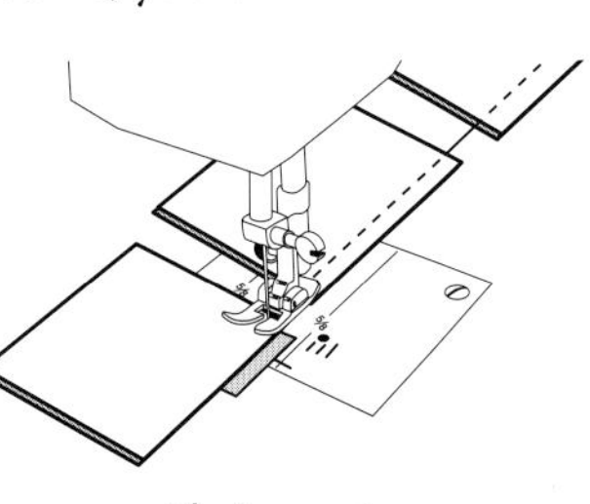

Chain sewing

Checking Your Work

As you stitch, it's a good idea to check the dimensions of the units you have sewn to make sure they are the correct size. It is much easier to correct problems at each stage of construction rather than later, when ripping seams and restitching become complicated and time-consuming. Interior pieces should measure the intended finished size, and outer pieces should measure the finished size plus ¼" all around for the last seam allowance. Check one last time when the blocks are complete. Rip and restitch units to size as necessary.

Borders

Straighten the edges of the quilt top before adding the borders. There should be little or no trimming needed for a straight-set quilt. A diagonally set quilt is often constructed with oversized side triangles that need to be trimmed to size. Align the ¼" line on the ruler with the block points and trim the quilt edges to ¼" from these points. Always position a ¼" line of the ruler along the block points of the adjacent edge at the same time, so that the corner will be square when the trimming is complete.

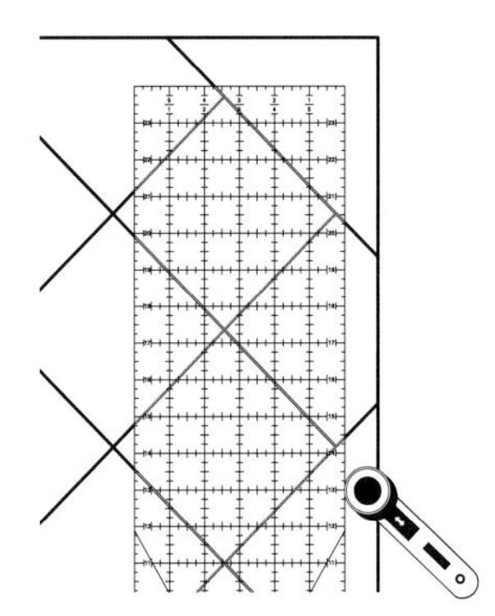

Trim the edges of the quilt to ¼" from the block points.

To find the correct measurement for plain border strips, always measure through the center of the quilt, not at the outside edges. This ensures that the borders are of equal length on opposite sides of the quilt and brings the outer edges into line with the center dimension if discrepancies exist. Otherwise, your quilt might not be "square," due to minor piecing errors and/or stretching that can occur while you work with the pieces.

1. Sew the border strips together, end to end, to make one continuous strip. Measure the quilt from the top to the bottom edge through the center of the quilt. From the long pieced strip, cut 2 border strips to this measurement and pin them to the sides of the quilt, easing to fit as necessary.

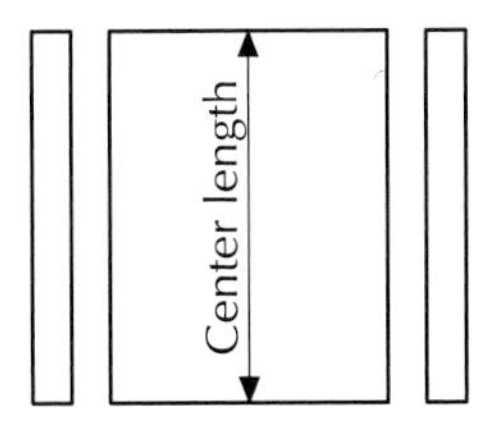

2. Sew the borders in place and press the seams toward the borders.
3. Measure the center width of the quilt, including the side borders, to determine the measurement of the top and bottom border strips. Cut the borders to this measurement, and pin them to the top and bottom of the quilt top, again easing to fit as necessary. Stitch in place and press the seams toward the border strips.

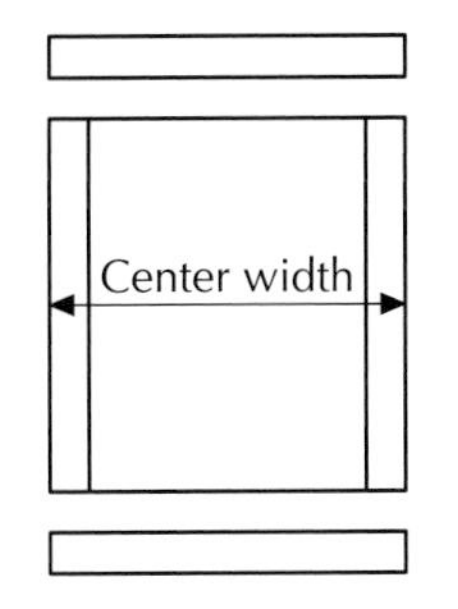

Quilt Backings

Yardage requirements are provided for making paneled quilt backings if the quilt is larger than 36" square. After washing the fabric, cut and sew it into a two- or three-panel backing as needed. Cut the backing 3" to 4" larger than the quilt top on all sides.

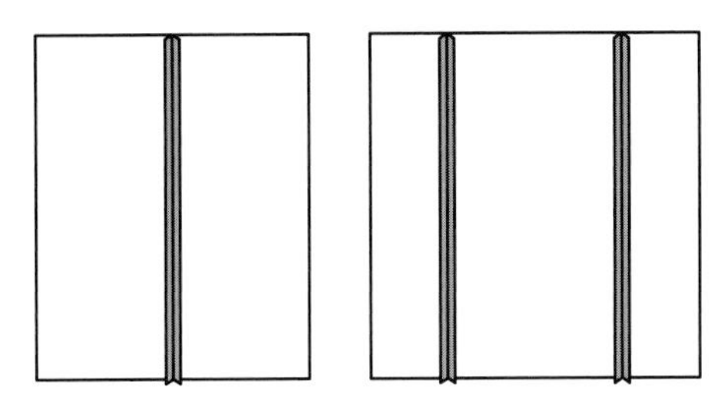

Two ways to piece paneled backings

Basting and Quilting

Basting the Layers

Mark the quilt top with the desired quilting designs and cut the batting 2" to 3" larger than the quilt top all the way around. Press the prepared backing smooth and tape it, right side down, to a clean, hard surface. Securely tape the sides to the surface every few inches. Tape the corners last. Smooth and center the batting over the backing, then carefully place the quilt top over the batting, right side up. Smooth out the quilt top and pin-baste the "quilt sandwich" through all three layers, always working from the center out.

Using a light-colored thread, thread-baste the sandwich in a 3" to 4" grid, again working from the center out. Baste across both diagonals. Secure the edges of the sandwich with a line of stitches around the edge. Remove the pins.

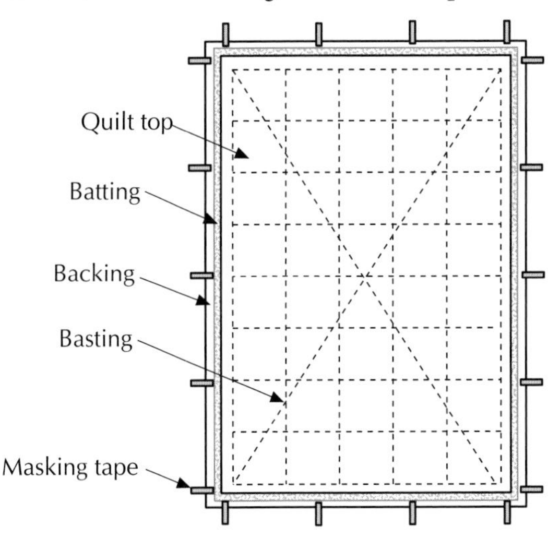

Many machine quilters and even some hand quilters prefer to use special quilter's safety pins rather than thread to baste the quilt sandwich. Place safety pins 4" to 6" apart and remove them as needed while stitching. Some quilters use plastic tacks, such as those used to secure price tags to clothing. They are available at quilt shops or through mail-order catalogs.

Quilting

Quilting is a simple running stitch that goes through all three layers of the quilt sandwich. Stitches are worked from the center out toward the edges. Most quilters prefer to use some kind of frame to keep the layers from shifting while quilting.

1. Thread a quilting needle (called a Between) with a 12" to 18" length of quilting thread. Make a single knot close to the end of the thread. Slip the needle through the quilt layers a needle's length from the starting point. Bring the needle up and give a small tug to lodge the knot in the layers.
2. Following the quilting marks, sew a simple running stitch through all 3 layers. This will take some practice to master.
3. End a line of quilting by making a small knot in the thread, about ⅛" from where it exits the quilt. Take the last stitch between the layers only and run the needle a short distance away from the last stitch before bringing the needle up and out of the quilt. Give a gentle tug, and the knot will slip between the layers. Clip the thread a short distance from the quilt top and work the end back between the layers.

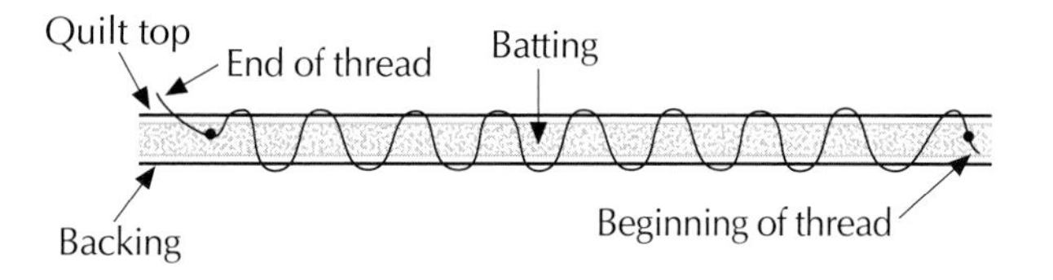

4. Quilt from the center out to the edges. Once all the quilting is done, remove the interior basting stitches, leaving the stitching around the edges of the quilt in place for binding.

Binding

The fabric requirements for the bindings in this book are based on straight-grain fabric strips for a double-fold binding.

1. Cut 2"-wide strips from selvage to selvage for standard ¼"-wide finished binding.
2. Join the strips at right angles and stitch across the corner to make one long piece of binding. Trim away excess fabric and press seams open. Use closely matching thread to avoid peekaboo stitches at the seams.

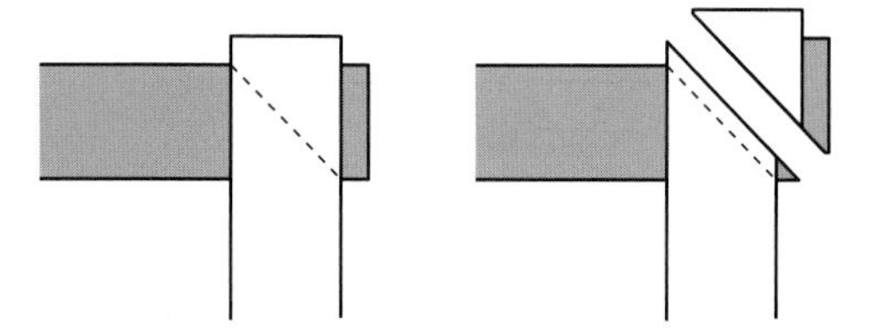

3. Fold the strip in half lengthwise, wrong sides together, and press. At one end of the strip, turn under ¼" at a 45° angle and press.

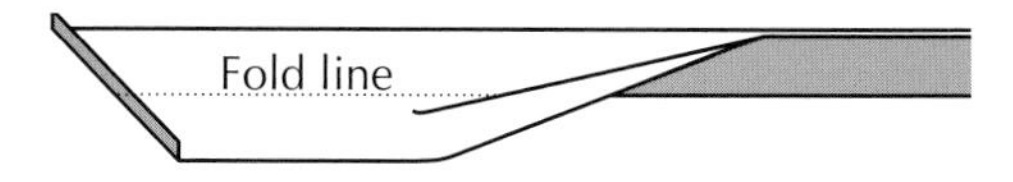

To attach the binding:

1. Baste the 3 layers of the quilt securely at the outer edges if you have not already done so.
2. Trim the batting and backing even with the quilt-top edges and corners.
3. In the center of one edge of the quilt, align the raw edges of the binding with the raw edge of the quilt top. Leaving about 6" free as a starting tail, sew the binding to the edge of the quilt with a ¼"-wide seam allowance. Stop stitching ¼" from the corner of the first side. Backstitch and remove the quilt from the machine.

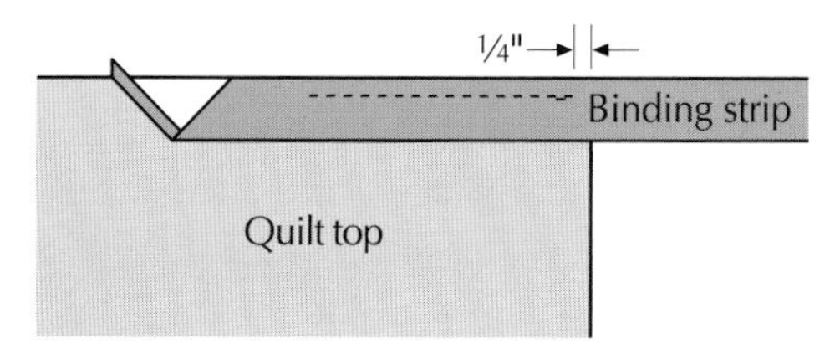

4. At the corner, flip the binding straight up from the corner so it forms a continuous line with the adjacent side of the quilt top.

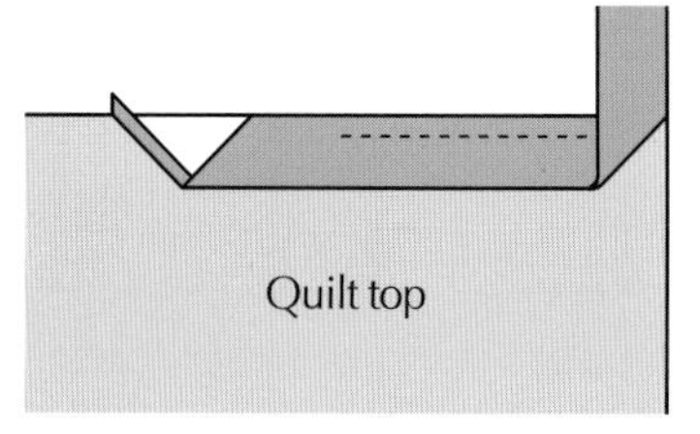

5. Fold the binding straight down so it lies on top of the next side. Pin the pleat in place. Starting at the edge, stitch the second side of the binding to the quilt, stopping at the ¼" mark on the next corner. Repeat for the remaining corners.

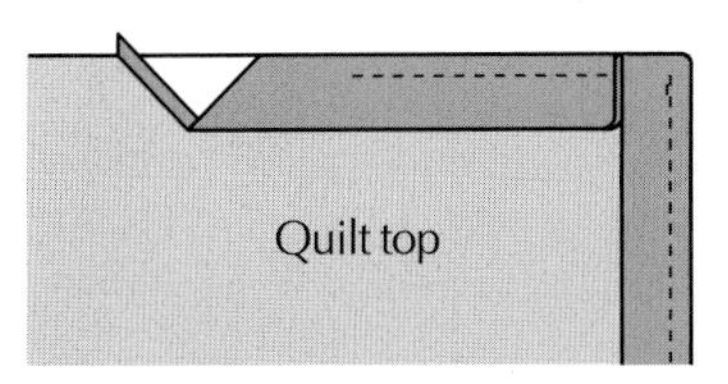

6. When you have turned the last corner and are nearing the point where you began, stop and overlap the binding by about 1". Cut away any excess binding, trimming the end to a 45° angle. Tuck the end into the fold and finish the seam.

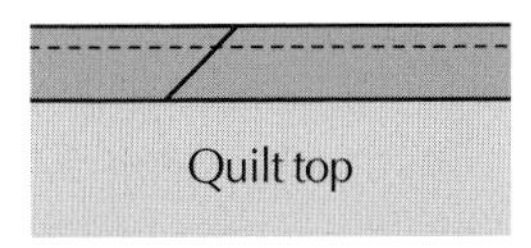

7. Turn the binding to the back of the quilt and slipstitch it to the backing to complete the binding—and your quilt!

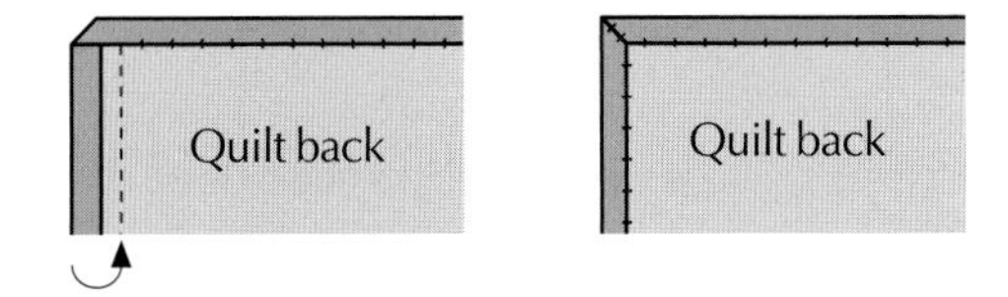

Labeling Your Quilt

Be sure to sign and date your quilt. Labels can be simple or elaborate. They can be handwritten, typed, or embroidered. Be sure to include the name of the quilt, your name, your city and state, the date, and the name of the recipient if it is a gift. Include interesting or important information about the quilt. Future generations will want to know more about the quilt than just who made it and when.

About the Author

Born and raised in southeastern Pennsylvania, Donna Lynn Thomas has had a needle in her hand since she was a little girl. Her mother was a home economics teacher and her father an engineer. It seems only natural that Donna, with a love for both fabric and geometry, would take to quilting.

Donna has been quilting since 1975 and teaching since 1982. The introduction of rotary-cutting tools in the 1980s revolutionized her approach to quiltmaking. Since Nancy J. Martin introduced her to bias strip piecing in 1987, Donna has worked exclusively with that method, developing new and innovative ways to maximize precision piecing. In 1995 Donna developed the Bias Stripper ruler to use with her bias strip-piecing methods.

Donna is the author of five other books: *Small Talk, A Perfect Match: A Guide to Precise Machine Piecing, Shortcuts to the Top, Stripples*, and *Stripples Strikes Again!*

The Thomas family includes Donna, her husband, Terry, and their two teenage sons, Joseph and Peter. Terry's military career took the family many places, giving Donna the opportunity to teach quilting all over the country as well as overseas. Terry's recent move to civilian life has given Donna visions of staying put in Pennsylvania, where she hopes to continue quilting, writing, and teaching as well as to pursue her other passion in life—gardening.